Tensor Calculus for Physics

Tensor Calculus for Physics

A Concise Guide

Dwight E. Neuenschwander

Dev Publishers & Distributors
New Delhi

Published by
Dev Publishers & Distributors
Second Floor, Prakashdeep,
22, Delhi Medical Association Road,
Darya Ganj, New Delhi - 110 002

www.devbooks.co.in

ISBN 978-93-94852-58-7

This edition, 2023

Printed in India.

To my parents, Dwight and Evonne

with gratitude

We were on a walk and somehow began to talk about space. I had just read Weyl's book Space, Time, and Matter, *and under its influence was proud to declare that space was simply the field of linear operations. "Nonsense," said Heisenberg, "space is blue and birds fly through it."* –Felix Bloch

Contents

Preface

By the standards of tensor experts, this book will be deemed informal, repetitious, and incomplete. But it was not written for those who are already experts. It was written for physics majors who are new to tensors and find themselves intrigued but frustrated by them.

The typical readers I have in mind are undergraduate physics majors in their junior or senior year. They have taken or are taking courses in classical mechanics and electricity and magnetism, have become acquainted with special relativity, and feel a growing interest in general relativity.

According to *Webster's Dictionary*, the word "concise" means brief and to the point. However, tensor initiates face a problem of so much being left unsaid. Few undergraduates have the opportunity to take a course dedicated to tensors. We pick up whatever fragments about them we can along the way. Tensor calculus textbooks are typically written in the precise but specialized jargon of mathematicians. For example, one tensor book "for physicists" opens with a discussion on "the group G_α and affine geometry." While appropriate as a logical approach serving the initiated, it hardly seems a welcoming invitation for drawing novices into the conversation. This book aims to make the conversation accessible to tensor novices.

An undergraduate who recently met the inertia and electric quadrupole tensors may feel eager to start on general relativity. However, upon opening some modern texts on the subject, our ambitious student encounters a new barrier in the language of differential forms. Definitions are offered, but to the novice the motivations that make those definitions worth developing are not apparent. One feels like having stepped into the middle of a conversation. So one falls back on older works that use "contravariant and covariant" language, even though newer books sometimes call such approaches "old-fashioned." Fashionable or not, at least they are compatible with a junior physics major's background and offer a useful place to start.

In this book we "speak tensors" in the vernacular. Chapter 1 reviews undergraduate vector calculus and should be familiar to my intended audience; I want to start from common ground. However, some issues taken for granted in familiar vector calculus are the tips of large icebergs. Chapter 1 thus contains both review and foreshadowing. Chapter 2 introduces tensors through scenarios encountered in an undergraduate physics curriculum. Chapters 3-6 further develop tensor calculus proper, including derivatives of tensors in spaces with curvature. Chapter 7 re-derives important tensor results through the use of basis vectors. Chapter 8 offers an informal introduction to differential forms, to show why these strange mathematical objects are beautiful and useful.

Jacob Bronowski wrote in *Science and Human Values*, "The poem or the discovery exists in two moments of vision: the moment of appreciation as much as that of creation.... We enact the creative act, and we ourselves make the discovery again." I thank the reader for accommpanying me as I retrace in these pages the steps of my own journey in coming to terms with tensors.

Acknowledgments

To the students in the undergraduate physics courses that I have been privileged to teach across the years I express my deep appreciation. Many of the arguments and examples that appear herein were first tried out on them. Their questions led to deeper investigations.

I thank the individuals who offered practical help during the writing of this book. Among these I would explicitly mention Nathan Adams, Brent Eskridge, Curtis McCully, Mohammad Niazi, Reza Niazi, Lee Turner, Kenneth Wantz, Johnnie Renee West, Mark Winslow, and Nicholas Zoller. I also thank the Catalysts, an organization of SNU science alumni, for their support and encouragement throughout this project.

Looking back over a longer timescale, I would like to acknowledge my unrepayable debt to all my teachers. Those who introduced me to the topics described herein include Raghunath Acharya, T.J. Bartlett, Steve Bronn, James Gibbons, John Gil, David Hestenes, John D. Logan, Richard Jacob, Ali Kyrala, George Rosensteel, Larry Weaver, and Sallie Watkins. Their influence goes deeper than mathematical physics. To paraphrase Will Durant, who acknowledged a mentor when introducing *The Story of Philosophy* (1926), may these mentors and friends find in these pages—incidental and imperfect though they are—something worthy of their generosity and their faith.

Many thanks to former Johns Hopkins University Press editor Trevor Lipscombe for initially suggesting this project, to JHUP editor Vincent Burke for continuing expert guidance, and to the ever-gracious JHUP editorial and production staff, including Catherine Goldstead, Hilary S. Jacqmin, Juliana McCarthy, and others behind the scenes.

Above all, with deep gratitude I thank my wife Rhonda for her support and patience, as this project—like most of my projects—took longer than expected.

- D. E. N

Tensor Calculus for Physics

Chapter 1

Tensors Need Context

1.1 Why Aren't Tensors Defined by What They *Are*?

When as an undergraduate student I first encountered the formal definition of a tensor, it left me intrigued but frustrated. It went something like this:

> *"A set of quantities $T^r{}_s$ associated with a point P are said to be the components of a second-order tensor if, under a change of coordinates, from a set of coordinates x^s to x'^s, they transform according to*
>
> $$T'^r{}_s = \frac{\partial x'^r}{\partial x^i}\frac{\partial x^j}{\partial x'^s}T^i{}_j \tag{1.1}$$
>
> *where the partial derivatives are evaluated at P."*

Say what?

Whatever profound reasons were responsible for saying it this way, this definition was more effective at raising questions than it was at offering answers. It did not define a tensor by telling me what it *is*, but claimed to define tensors by describing how they *change*. What kind of definition is that supposed to be, that doesn't tell you what it is that's changing? It seemed like someone defining "money" by reciting the exchange rate between dollars and Deutchmarks—however technically correct the statement might be, it would not be very enlightening to a person encountering the concept of money for the first time in their lives.

Furthermore, why was the change expressed in terms of a coordinate transformation? If $T^r{}_s$ is a component, what is the *tensor itself*? Why were some indices written as superscripts and others as subscripts? Why were some indices repeated? What does the "order" of a tensor mean (also called the tensor's "rank" by some authors)?

Such a definition was clearly intended for readers already in command of a much broader mathematical perspective than I had at the time. Treatises that begin with such formal definitions aim to develop the logical structure of tensor calculus with rigor. That approach deserves the highest respect. But it is an acquired taste. Despite its merits, that will not be the approach taken here. For instance, I will not use phrases such as "to every subspace V_r of E_n there corresponds a unique complementary subspace W_{n-r}." Rather, we will first encounter a few specific tensors as they emerge in physics applications, including classical mechanics and electrodynamics. Such examples will hopefully provide context where motivations behind the formal definition will become apparent.

In the meantime, however, we have been put on notice that coordinate transformations are evidently an essential part of what it means to be a tensor. What are coordinates? They are maps introduced by us to solve a particular problem. What are coordinate transformations? If I use one coordinate system and you use a different one to map Paris (or the hydrogen atom) a transformation provides the dictionary for converting the address of the Louvre (or of the electron) from my coordinate system to yours. This elementary fact suggests that the tensor definition cited above forms a theory of relativity, because any theory of relativity describes how quantities and relationships are affected under a change of reference frame.

A coordinate system is chosen as a matter of taste and convenience, a map *we create* to make solving a problem as simple as possible. Maps are tools to help us make sense of reality; they are not reality itself. Just as the physical geography of the Earth does not depend on whether the mapmaker puts north at the top or at the bottom of a map, likewise the principles and conclusions of physics must transcend a choice of coordinates. Reconciling these competing values–of choosing coordinates for convenience but not being dependent on the choice–lies at the heart of the study of tensors.

Any theory of relativity is built on a foundation of what stays the same under a change of reference frame. Quantities that stay the same are called "invariants" or "scalars." For example, the distance between two points in Euclidean space does not depend on how we orient the coordinate axes: length is invariant, a scalar. To say that a number λ is a scalar means that when we transform from one system of coordinates to another, say, from coordinates (x,y,z) to (x', y', z'), λ remains unaffected by the transformation:

$$\lambda' = \lambda. \tag{1.2}$$

This statement formally *defines* what it means for a quantity to be a scalar.

Notice how this formal definition of scalars describes how they transform (in particular, scalars stay the same) under a change of reference frame. Suggestively, it resembles the statement encountered above in the formal definition of a tensor. Although operational definitions prescribe how a body's mass and temperature can be measured, identifying mass and temperature *as scalars* has no relevance until a change of coordinate system becomes an issue.

In contrast to scalars, vectors are introduced in elementary treatments as "arrows" that carry information about a magnitude and a direction. That information is independent of any coordinate system. But when a coordinate system is introduced, a vector can be partitioned into a set of numbers, the "components," whose separate numerical values depend on the orientation and scaling of the coordinate axes. Because coordinate systems can be changed, equations that include vector components must inevitably be examined in the light of their behavior under coordinate transformations. Vectors offer a crucial intermediate step from scalars to the tensors that belong to the formal definition introduced above.

Some books launch into tensor analysis by making a distinction between vectors and something else called their "duals" or "one-forms" (with extensions to "two-forms" and so on). Feeling like someone stepping cluelessly into the middle of a conversation, one gathers that dual vectors are related to vectors, but are not necessarily identical to vectors—although sometimes they are! This raises challenges to mustering motivation. Upon being informed that a one-form is a way to slice up space, the novice is left to wonder, "Why would anyone want to consider such a thing? Why is this necessary? What am I missing?" This problem is especially acute if the reader is pursuing an interest in tensor calculus through self-study, without ready access to mentors who are already experts, who could fill in the gaps that remain unexplained in textbook presentations.

Let us look backward at how vectors are discussed in introductory physics. We will first do so without coordinates and then introduce coordinates as a subsequent step. Then we can look forward to appreciating scalars and vectors as special cases of tensors.

1.2 Euclidean Vectors, without Coordinates

Introductory physics teaches that "vectors are quantities having magnitude and direction." *Direction* seems to be the signature characteristic of vectors which distinguishes them from mere numbers. Let us indulge in a brief review of the elementary but crucial lessons about vectors and their doings which we learned in introductory and intermediate physics. Much about tensors is foreshadowed by, and grows from, vectors.

Perhaps the first lesson in vector education occurs when one learns how to draw an arrow from point A to point B—a "displacement"—and declares that the arrow represents "the vector from A to B." Vectors are commonly denoted typographically with boldface fonts, such as **V**, or with an arrow overhead, such as $\vec{V}$. Any vector carries a magntiude and a direction, making it necessary to emphasize the distinction between the vector **V** and its magnitude $|\mathbf{V}| \equiv V$.

To make a set of vectors into a mathematical system it is necessary to define rules for how they may be combined. The simplest vector operations are "parallelogram addition" and "scalar multiplication."

Imagine the arrow representing a vector, hanging out there in space all by itself. We can move it about in space while maintaining its magnitude and direction, an operation called "parallel transport." Parallelogram addition defines a vector sum $\mathbf{V}+\mathbf{W}$ operationally: parallel-transport $\mathbf{W}$ so that its tail joins to the head of $\mathbf{V}$, and then draw the "resultant" from the tail of $\mathbf{V}$ to the head of $\mathbf{W}$. Parallelogram addition of two vectors yields another vector.

If α is a scalar and $\mathbf{V}$ a vector, then $\alpha\mathbf{V}$ is another vector, rescaled compared to $\mathbf{V}$ by the factor α, pointing in the same direction as $\mathbf{V}$ if $\alpha > 0$, but in the reverse direction if $\alpha < 0$. In particular, $(-1)\mathbf{V} \equiv -\mathbf{V}$ is a vector that has the same magnitude as $\mathbf{V}$ but points in the opposite direction. A scalar can be a pure number like 3 or 2π, or it might carry dimensions such as mass or electric charge. For instance, in elementary Newtonian mechanics, linear momentum $\mathbf{p}$ is a rescaled velocity, $\mathbf{p} = m\mathbf{v}$.

Vector subtraction is defined, not as another operation distinct from addition, but in terms of scalar multiplication and parallelogram addition, because $\mathbf{A}-\mathbf{B}$ means $\mathbf{A}+(-1)\mathbf{B}$. The zero vector $\mathbf{0}$ can be defined, such that $\mathbf{A}+\mathbf{0} = \mathbf{A}$, or equivalently $\mathbf{A}+(-1)\mathbf{A} = \mathbf{0}$. Geometrically, $\mathbf{0}$ has zero magnitude and no direction can be defined for it, because its tail and head are the same point.

A dimensionless vector $\hat{\mathbf{V}}$ of unit magnitude that points in the same direction as $\mathbf{V}$ is constructed from $\mathbf{V}$ by forming $\frac{\mathbf{V}}{|\mathbf{V}|} \equiv \hat{\mathbf{V}}$. Said another way, any vector $\mathbf{V}$ can be factored into the product of its magnitude and direction, according to $\mathbf{V} = V\hat{\mathbf{V}}$. In a space of two or more independent directions, having at one's disposal a unit vector for each dimension makes it possible to build up any vector through the scalar multiplication of the unit vectors, followed by parallelogram addition. We will return to this notion when we discuss vectors with coordinates.

A more sophisticated algebra can be invented by introducing vector multiplications. When you and I labor to push a piano up a ramp, the forces parallel to the piano's displacement determine its acceleration along the ramp. For such applications the "dot product" falls readily to hand, defined without coordinates according to

$$\mathbf{A}\cdot\mathbf{B} \equiv |\mathbf{A}||\mathbf{B}|\cos\theta, \tag{1.3}$$

where θ denotes the angle between $\mathbf{A}$ and $\mathbf{B}$. The number that results from the dot product loses all information about direction. If the dot product were a machine with two input slots into which you insert two vectors, the output is a scalar. To be rigorous, we need to prove this claim that the dot product of two vectors yields a scalar. We will turn to that task when we discuss theories of relativity. Anticipating that result, the dot product is also called the scalar product. As an invariant, the scalar product plays a central role in geometry and in theories of relativity, as we shall see.

Another definition of vector multiplication, where the product of two vectors gives another vector, finds hands-on motivation when using a wrench to loosen a bolt. The force must be applied *off* the bolt's axis. If the bolt is too tight, you can apply more force—*or* get a longer wrench. To describe quantitatively what's going on, visualize the lever arm as a vector $\mathbf{r}$ extending from the bolt's axis to the point on the wrench where the force is applied. That lever arm and force produce a torque, an example of a new vector defined in terms of the old ones by the "cross product." In coordinate-free language, and in three dimensions, the cross product between two vectors $\mathbf{A}$ and $\mathbf{B}$ is defined as

$$\mathbf{A} \times \mathbf{B} \equiv (|\mathbf{A}||\mathbf{B}| \sin \theta)\hat{\mathbf{n}} \tag{1.4}$$

where the direction of the unit vector $\hat{\mathbf{n}}$ is given by the right-hand rule: point the fingers of your right hand in the direction of $\mathbf{A}$, then turn your hand towards $\mathbf{B}$. Your thumb points in the direction of $\hat{\mathbf{n}}$. The result, $\mathbf{A} \times \mathbf{B}$, is perpendicular to the plane defined by the two vectors $\mathbf{A}$ and $\mathbf{B}$. Because the cross product of two vectors yields another vector (an assertion that rigorously must be proved), it is also called the vector product.

Notice that the dot product is symmetric, $\mathbf{A} \cdot \mathbf{B} = \mathbf{B} \cdot \mathbf{A}$, but the cross product is antisymmetric, $\mathbf{A} \times \mathbf{B} = -\mathbf{B} \times \mathbf{A}$.

1.3 Derivatives of Euclidean Vectors with Respect to a Scalar

With vector subtraction and scalar multiplication defined, all the necessary ingredients are in place to define the derivative of a vector with respect to a scalar. Instantaneous velocity offers a prototypical example. First, record a particle's location at time t. That location is described by a vector $\mathbf{r}(t)$, which points from some reference location—the displacement from an origin—to the particle's instantaneous location. Next, record the particle's location again at a time $t + \Delta t$ later, expressed as the vector $\mathbf{r}(t + \Delta t)$. Subtract the two vectors, divide by Δt (in scalar multiplication language, dividing by Δt means multiplication by the scalar $1/\Delta t$), and then take the limit as Δt approaches zero. The result is the instantaneous velocity vector $\mathbf{v}(t)$:

$$\mathbf{v}(t) \equiv \lim_{\Delta t \to 0} \frac{\mathbf{r}(t + \Delta t) - \mathbf{r}(t)}{\Delta t} \equiv \frac{d\mathbf{r}}{dt}. \tag{1.5}$$

As parallelogram addition shows, the displacement vector $d\mathbf{r}$ and thus the velocity vector $\mathbf{v}$ are tangent to the trajectory swept out by $\mathbf{r}(t)$.

To have a self-contained mathematical system in the algebra and calculus of a set of vectors, operations between elements of a set must produce another element within the same set, a feature called "closure." As a tangent vector, does $\mathbf{v}$ reside in the same space as the original vector $\mathbf{r}(t)$ used to derive it? For projectile problems analyzed in a plane, the velocity vector lies in the same plane as the instantaneous position vector. Likewise, when a particle's motion sweeps out a trajectory in three-dimensional Euclidean space, the velocity vector exists in that same three-dimensional space. In all cases considered in introductory physics, the derivative of a vector resides in the same space as the vector being differentiated. The circumstances under which this happens should not be taken for granted. They will occupy our attention in Chapter 4 and again in Chapter 7, but a word of foreshadowing can be mentioned here.

In introductory physics, we are used to thinking of vectors as a displacement *within* Euclidean space. That Euclidean space is often mapped with an xyz coordinate system. Other systems are later introduced, such as cylindrical and spherical coordinates (see Appendix A), but they map the same space. When we say that the particle is located at a point given by the position vector $\mathbf{r}$ relative to the origin, we are saying that $\mathbf{r}$ is the displacement vector from the origin to the particle's location,

$$\mathbf{r} = (x, y, z) - (0, 0, 0). \tag{1.6}$$

When velocity vectors come along, they still reside in the same xyz system as does the original position vector $\mathbf{r}$. The space of tangent vectors, or "tangent space," happens to be identical to the original space. But as Chapter 7 will make explicit, vectors defined by such displacements reside formally only in the *local* tangent space. To visualize what this means, imagine a vector that points, say, from the base of a flagpole on your campus to the front door of the physics building. That is a displacement vector, pointing from one point to another on a patch of surface sufficiently small, compared to the entire planet's surface, that your campus can be mapped with a Euclidean xy plane. But if you extend that vector's line of action to infinity, the Earth's surface curves out from beneath it. Therefore, other than on a locally flat patch of surface, the vector does not exist on the globe's surface.

From the derivative of a displacement with respect to a scalar, we turn to the derivative of a scalar with respect to a displacement and meet the gradient.

1.4 The Euclidean Gradient

Examples of scalar fields encountered in everyday life include atmospheric pressure and temperature. They are *fields* because they are functions of location in space and of time. The gradient operator takes the derivative of a scalar function and produces a vector. Whereas velocity was a coordinate displacement divided by the change in a scalar, the gradient is the change in a scalar divided by a displacement. For that reason, the velocity and the gradient are examples of families of vectors that are said to be "reciprocal" or "dual" to one another–a distinction we do not need for now, but will develop in Chapter 3 and revisit from another perspective in Chapter 7.

In three-dimensional Euclidean space, the gradient of a scalar function ϕ is denoted $\nabla\phi$. Together with a unit vector $\hat{\mathbf{n}}$, it defines the coordinate-free "directional derivative" $(\boldsymbol{\nabla}\phi)\cdot\hat{\mathbf{n}}$. To interpret it, let us denote as $\frac{\partial\phi}{\partial n}$ the instantaneous slope of the function ϕ in the direction of $\hat{\mathbf{n}}$. Imagine standing on a mountainside. Walking east, one may find the slope to be zero (following a contour of constant elevation), but stepping south one may tumble down a steep incline! From the coordinate-free definition of the scalar product, it follows that

$$\frac{\partial\phi}{\partial n} = |\boldsymbol{\nabla}\phi|\cos\theta, \tag{1.7}$$

where θ denotes the angle between $\boldsymbol{\nabla}\phi$ and $\hat{\mathbf{n}}$. The slope, $\partial\phi/\partial n$, will be greatest when $\cos\theta = 1$, when the gradient and $\hat{\mathbf{n}}$ point in the same direction. In other words, the gradient of a scalar function ϕ is the vector that points in the direction of the steepest ascent of ϕ at the location where the derivative is evaluated. The gradient measures the change in ϕ *per unit length.*

The gradient, understood as a vector in three-dimensional space, can be used in the dot and cross products, which introduces two first derivatives of vectors with respect to spatial displacements: the divergence and the curl. The "divergence" of a vector field is defined as $\boldsymbol{\nabla}\cdot\mathbf{A}$, which is nonzero if streamlines of $\mathbf{A}$ diverge away from the point where the scalar product is evaluated. The "curl" of a vector field, $\boldsymbol{\nabla}\times\mathbf{A}$, is nonzero if the streamlines of $\mathbf{A}$ form whirlpools in the neighborhood of the point where the curl is evaluated.

Second derivatives involving $\boldsymbol{\nabla}$ which we shall meet repeatedly include the Laplacian $\boldsymbol{\nabla}\cdot(\boldsymbol{\nabla}\phi) = \nabla^2\phi$, the identities $\boldsymbol{\nabla}\times(\boldsymbol{\nabla}\phi) = \mathbf{0}$ for any scalar ϕ, and $\boldsymbol{\nabla}\cdot(\boldsymbol{\nabla}\times\mathbf{B}) = 0$ for any vector $\mathbf{B}$.

1.5 Euclidean Vectors, with Coordinates

The story goes that early one morning in 1637, René Descartes was lazing in bed and happened to notice a fly walking on the wall. It occured to Descartes that the fly's location on the wall at any instant could be specified by two numbers: its distance vertically above the floor, and its horizontal distance from a corner of the room. As the fly ambled across the wall, these numbers would change. The set of all the pairs of numbers would map out the fly's path on the wall. Thus goes the advent story of the Cartesian coordinate system, the familiar (x, y) grid for mapping locations in a plane. When the fly leaps off the wall and buzzes across the room, its flight trajectory can be mapped in three dimensions with the ordered triplet (x, y, z), where each coordinate is a function of time.

In three-dimensional Euclidean space with its three mutually perpendicular axes and their (x, y, z) coordinates, it is customary to define three dimensionless unit vectors: $\hat{\mathbf{i}}$, $\hat{\mathbf{j}}$, and $\hat{\mathbf{k}}$, which point respectively in the direction of increasing x, y, and z. As unit vectors that are mutually perpendicular, or "orthogonal," their dot products are especially simple. Being unit vectors,

$$\hat{\mathbf{i}}\cdot\hat{\mathbf{i}} = \hat{\mathbf{j}}\cdot\hat{\mathbf{j}} = \hat{\mathbf{k}}\cdot\hat{\mathbf{k}} = 1, \tag{1.8}$$

and being mutually orthogonal,

$$\hat{\mathbf{i}} \cdot \hat{\mathbf{j}} = \hat{\mathbf{i}} \cdot \hat{\mathbf{k}} = \hat{\mathbf{j}} \cdot \hat{\mathbf{k}} = 0. \tag{1.9}$$

We can compress these six relations into one line by first denoting $\hat{\mathbf{i}} \equiv \hat{\mathbf{e}}_1, \hat{\mathbf{j}} \equiv \hat{\mathbf{e}}_2, \hat{\mathbf{k}} \equiv \hat{\mathbf{e}}_3$, and summarize all these relations with

$$\hat{\mathbf{e}}_n \cdot \hat{\mathbf{e}}_m = \delta_{nm}, \tag{1.10}$$

where $n, m = 1, 2, 3$ and we have introduced the "Kronecker delta" symbol δ_{nm}, which equals 1 if $n = m$ and equals 0 if $n \neq m$. Being orthogonal and "normalized" to unit magnitude, these vectors are said to be "orthonormal." Any vector in the space can be written as a superposition of them, thanks to scalar multiplication and parallelogram addition. For example, consider the position vector $\mathbf{r}$:

$$\mathbf{r} = x\hat{\mathbf{e}}_1 + y\hat{\mathbf{e}}_2 + z\hat{\mathbf{e}}_3. \tag{1.11}$$

To make such sums more compact, let us denote each component of $\mathbf{r}$ with a superscript: $x = x^1, y = x^2, z = x^3$. We use superscripts so that our discussion's notation will be consistent with other literature on tensor calculus. Books that confine their attention to Euclidean spaces typically use subscripts for coordinates (e.g., Ch. 1 of Marion and Thornton), such as $x = x_1$ and $y = x_2$. But our subject will take us beyond rectangular coordinates, beyond three dimensions, and beyond Euclidean spaces. In those domains some vectors behave like displacement vectors, and their components are traditionally denoted with superscripts. But other vectors exist, which behave somewhat differently, and their components are denoted with subscripts. In Euclidean space this distinction does not matter, and it becomes a matter of taste whether one uses superscripts or subscripts. However, so that we will not have to redefine component notation later, we will use superscripts for vector components from the outset. Whenever confusion might result between exponents and superscripts, parentheses will be used, so that $(A^1)^2$ means the square of the x-component of $\mathbf{A}$. Now any vector $\mathbf{A}$ may be written in terms of components and unit vectors as

$$\mathbf{A} = A^1\hat{\mathbf{i}} + A^2\hat{\mathbf{j}} + A^3\hat{\mathbf{k}} \tag{1.12}$$

or more compactly as

$$\mathbf{A} = \sum_{i=1}^{3} A^i \hat{\mathbf{e}}_i. \tag{1.13}$$

A set of vectors, in terms of which all other vectors in the space can be expressed by superposition, are called a "basis" if they are "linearly independent" and "span the space." To span the space means that *any* vector $\mathbf{A}$ in the space can be written as a superposition of the basis vectors, as in Eq. (1.12). These basis vectors $\hat{\mathbf{i}}$, $\hat{\mathbf{j}}$, and $\hat{\mathbf{k}}$ happen to be normalized to be dimensionless and of unit length, but being *unit* vectors, while convenient, is not essential to being a *basis* set. What *is* essential for a set of vectors to serve as a basis is their being *linearly independent.* To be linearly independent means that none of the basis vectors can be written as a superposition of the others. Said another way, $\mathbf{u}, \mathbf{v}, \mathbf{w}$ are linearly independent if and only if the equation

$$a\mathbf{u} + b\mathbf{v} + c\mathbf{w} = \mathbf{0} \tag{1.14}$$

requires $a = b = c = 0$.

With orthonormal unit basis vectors one "synthesizes" the vector $\mathbf{A}$ by choosing components A^x, A^y, and A^z and with them constructs the sum of Eq. (1.12). Conversely, if $\mathbf{A}$ is already given, its components can be determined in terms of the unit basis vectors in a procedure called "analysis," where, thanks to the orthonormality of the basis vectors,

$$A^x = \mathbf{A} \cdot \hat{\mathbf{i}}, \tag{1.15}$$

$$A^y = \mathbf{A} \cdot \hat{\mathbf{j}}, \tag{1.16}$$

$$A^z = \mathbf{A} \cdot \hat{\mathbf{k}}. \tag{1.17}$$

An orthonormal basis is also said to be "complete." We will shortly have an efficient way to summarize all this business in a "completeness relation," when we employ a notation introduced by Paul Dirac.

Let $\mathbf{e}_n$ or $\vec{e}_n$ denote a basis vector; the subscript identifies the basis vector, not the component of a vector. Notice the notational distinction between basis vectors not necessarily normalized to unity, $\vec{e}_n$ (with arrows), on the one hand, and *unit* basis vectors such as $\hat{\mathbf{e}}_n$ (boldface with hat) on the other hand. If the basis vectors are not unit vectors, Eq. (1.10) gets replaced by

$$\vec{e}_n \cdot \vec{e}_m = g_{nm}, \tag{1.18}$$

where the g_{nm} are a set of coefficients we will meet again in Chapter 3 and thereafter, where they are called the coefficients of the "metric tensor." In this case the scalar product between two vectors $\mathbf{A} = A^1\vec{e}_1 + A^2\vec{e}_2 + A^3\vec{e}_3$ and a vector $\mathbf{B}$ similarly expressed takes the form

$$\begin{aligned} \mathbf{A} \cdot \mathbf{B} &= \sum_{n=1}^{3}\sum_{m=1}^{3} (A^n\vec{e}_n) \cdot (B^m\vec{e}_m) \\ &= \sum_{n=1}^{3}\sum_{m=1}^{3} (\vec{e}_n \cdot \vec{e}_m)A^n B^m \\ &= \sum_{n=1}^{3}\sum_{m=1}^{3} g_{nm}A^n B^m. \end{aligned}$$

Until Chapter 7 all our basis vectors will be unit vectors. Then in Chapter 7 we will make use of Eq. (1.18) again. For now I merely wanted to alert you to this distinction between matters of principle and matters of convenience.

If one set of axes gets replaced with another, the components of a given vector will, in the new system, have different values compared to its components in the first system. For example, a Cartesian system of coordinate axes can be replaced with spherical coordinates (r, θ, φ) and their corresponding unit basis vectors $\hat{\mathbf{r}}$, $\hat{\boldsymbol{\theta}}$, and $\hat{\boldsymbol{\varphi}}$; or by cylindrical coordinates (ρ, φ, z) with their basis vectors $\hat{\boldsymbol{\rho}}$, $\hat{\boldsymbol{\varphi}}$, and $\hat{\mathbf{z}}$ (see Appendix A). A set of axes can be rotated to produce a new set with a different orientation. Axes can be rescaled or inverted. One reference frame might move relative to another. However, the meaning of a vector—the information it represents about direction and magnitude—transcends the choice of coordinate axes. Just as the existence of the Louvre does not depend on its location being plotted on someone's map, a vector's existence does not depend on the existence of coordinate axes.

As a quantity carrying information about magnitude and direction, when a vector is written in terms of coordinate axes, the burden of carrying that information shifts to the components. The number of components must equal the number of dimensions in the space. For example, in the two-dimensional Euclidean plane mapped with xy coordinates, thanks to the theorem of Pythagoras the magnitude of $\mathbf{A}$, in terms of its components, equals

$$|\mathbf{A}| = \sqrt{(A^x)^2 + (A^y)^2}, \tag{1.19}$$

and the direction of $\mathbf{A}$ relative to the positive x-axis is $\theta = \tan^{-1}(\frac{A^y}{A^x})$. Conversely, if we know $|\mathbf{A}|$ and θ, then the components are given by $A^x = \mathbf{A} \cdot \hat{\mathbf{i}} = |\mathbf{A}| \cos\theta$ and $A^y = \mathbf{A} \cdot \hat{\mathbf{j}} = |\mathbf{A}| \sin\theta$.

In a system of rectangular coordinates, with vectors $\mathbf{A}$ and $\mathbf{B}$ written in terms of components on an orthonormal basis,

$$\mathbf{A} = A^x\hat{\mathbf{i}} + A^y\hat{\mathbf{j}} + A^z\hat{\mathbf{k}} \tag{1.20}$$

$$\mathbf{B} = B^x\hat{\mathbf{i}} + B^y\hat{\mathbf{j}} + B^z\hat{\mathbf{k}}, \tag{1.21}$$

their scalar product is defined, in terms of these components, as

$$\begin{aligned} \mathbf{A} \cdot \mathbf{B} &\equiv A^x B^x + A^y B^y + A^z B^z \\ &= \sum_{i=1}^{3} A^i B^i. \end{aligned}$$

We can see how this definition of the scalar product of $\mathbf{A}$ and $\mathbf{B}$ follows from the orthonormality of the basis vectors:

$$\begin{aligned} \mathbf{A} \cdot \mathbf{B} &= \sum_{n=1}^{3} \sum_{m=1}^{3} (A^n \hat{\mathbf{e}}_n) \cdot (B^m \hat{\mathbf{e}}_m) \\ &= \sum_{n=1}^{3} \sum_{m=1}^{3} A^n B^m \hat{\mathbf{e}}_n \cdot \hat{\mathbf{e}}_m \\ &= \sum_{n=1}^{3} \sum_{m=1}^{3} A^n B^m \delta_{nm} \\ &= \sum_{n=1}^{3} A^n B^n. \end{aligned}$$

With trig identities and direction cosines this definition of the scalar product in terms of components can be shown to be equivalent to $AB\cos\theta$.

The vector product $\mathbf{A}\times\mathbf{B}$ is customarily defined in terms of rectangular coordinates by the determinant

$$\mathbf{A}\times\mathbf{B} \equiv \begin{vmatrix} \hat{\mathbf{i}} & \hat{\mathbf{j}} & \hat{\mathbf{k}} \\ A^x & A^y & A^z \\ B^x & B^y & B^z \end{vmatrix}. \tag{1.22}$$

The equivalence of this definition with $\hat{\mathbf{n}}(AB\sin\theta)$ is left as an exercise.

In Cartesian coordinates, the ith component of the vector product may also be written

$$(\mathbf{A}\times\mathbf{B})^i = \sum_{k=1}^{3}\sum_{j=1}^{3}\varepsilon^{ijk}A^jB^k, \tag{1.23}$$

where by definition the "Levi-Civita symbol" $\varepsilon^{ijk} = +1$ if ijk forms an even permutation of the three-digit sequence $\{123\}$, $\varepsilon^{ijk} = -1$ if ijk forms an odd permutation of $\{123\}$, and $\varepsilon^{ijk} = 0$ if any two indices are equal. Permutations are generated by switching an adjacent pair; for example, to get 312 from 123, first interchange the 2 and 3 to get 132. A second permutation gives 312. Thus, $\varepsilon^{132} = -1$ and $\varepsilon^{312} = +1$, but $\varepsilon^{122} = 0$.

1.6 Euclidean Vector Operations with and without Coordinates

Coordinate systems are not part of nature. They are maps introduced by us for our convenience. The essential relationships in physics—which attempt to say something real about nature—must therefore transcend the choice of coordinates. Whenever we use coordinates, it must be possible to change freely from one system to another without losing the essence of the relationships that the equations are supposed to express. This puts constraints on what happens to vector components when coordinate systems are changed.

This principle of coordinate transcendence can be illustrated with an analogy from geometry. A circle of radius R is defined as the set of all points in a plane located at a fixed distance from a designated point, the center. This *concept* of the circle of radius R needs no coordinates. However, the *equation* of a circle looks different in different coordinate systems. In polar coordinates the circle centered on the origin is described by the simple equation $r = R$, and in Cartesian coordinates the same circle is described by the more complicated equation $\sqrt{x^2+y^2} = R$. The equations look different, but they carry identical information.

For a physics example of a coordinate-transcending relation between vectors, consider the relation between a conservative force $\mathbf{F}$ and its corresponding potential energy U. The force is the negative gradient of the potential energy, $\mathbf{F} = -\boldsymbol{\nabla} U$. This means that the force $\mathbf{F}$ points in the direction of the steepest decrease in U. Coordinates are not necessary to describe this relationship. But when we sit down to calculate the gradient of a given potential energy function, we typically find it expressed in rectangular, spherical, or cylindrical coordinates. The gradient looks very different in these coordinates, even though all three ways of writing it describe the same interaction. For instance, in rectangular coordinates the gradient looks like this:

$$\boldsymbol{\nabla} U = \hat{\mathbf{i}}\frac{\partial U}{\partial x} + \hat{\mathbf{j}}\frac{\partial U}{\partial y} + \hat{\mathbf{k}}\frac{\partial U}{\partial z}. \tag{1.24}$$

In cylindrical coordinates it takes the form

$$\boldsymbol{\nabla} U = \hat{\boldsymbol{\rho}}\frac{\partial U}{\partial \rho} + \hat{\boldsymbol{\varphi}}\frac{1}{\rho}\frac{\partial U}{\partial \varphi} + \hat{\mathbf{z}}\frac{\partial U}{\partial z}, \tag{1.25}$$

while in spherical coordinates the gradient gets expressed as

$$\boldsymbol{\nabla} U = \hat{\mathbf{r}}\frac{\partial U}{\partial r} + \hat{\boldsymbol{\theta}}\frac{1}{r}\frac{\partial U}{\partial \theta} + \hat{\boldsymbol{\varphi}}\frac{1}{r\sin\theta}\frac{\partial U}{\partial \varphi}. \tag{1.26}$$

All of these equations say the same thing: $\mathbf{F} = -\boldsymbol{\nabla} U$. Thus, when the expression gets presented to us in one coordinate system, as a matter of principle it must be readily translatable into any other system while leaving the information content intact.

As a worked example of a physics calculation transcending coordinates, consider a scenario familiar from elementary mechanics: Newton's laws applied to the ubiquitious block on the inclined plane. We model the block as a particle of mass m, model its interactions with the rest of the world through the language of force vectors, and choose to do physics in an unaccelerated reference frame, where the working equation of Newtonian mechanics is Newton's second law applied to the particle,

$$\mathbf{F} = m\mathbf{a}. \tag{1.27}$$

$\mathbf{F}$ denotes the vector sum of all the forces, and $\mathbf{a}$ denotes the particle's acceleration in response to $\mathbf{F}$.

The dominant forces on the block are those of gravity $m\mathbf{g}$ exerted by the Earth and contact forces exerted on the block by the plane, which includes friction $\mathbf{f}$ tangent to the block's surface, and the normal force $\mathbf{N}$. When we write out what $\mathbf{F}$ means in this situation, $\mathbf{F} = m\mathbf{a}$ becomes

$$m\mathbf{g} + \mathbf{N} + \mathbf{f} = m\mathbf{a}. \tag{1.28}$$

This statement as it stands contains the physics. However, to use this vector equation to solve for one or more unknowns, for example, to find the inclined plane's angle that will result in an acceleration down the plane of 3.5 m/s^2 when the coefficient of friction is 0.2, it is convenient to make use of parallelogram addition in reverse and project the vector equation onto a set of coordinate axes.

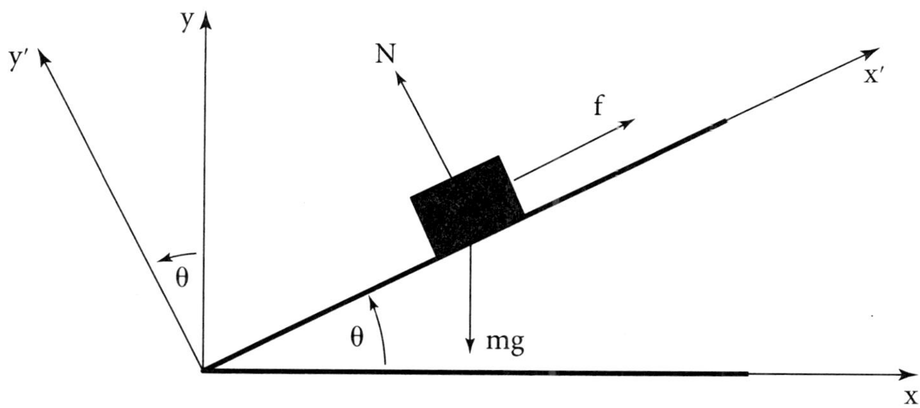

Figure 1.1: *The two coordinate axes xy and $x'y'$*

A set of mutually perpendicular axes can be oriented an infinite number of possible ways, including the (x, y) and $(x'y')$ axes shown in Fig. 1.1. The coordinate-free $\mathbf{F} = m\mathbf{a}$ can now be projected onto either set of axes. Formally, we use the scalar product to multiply Newton's second law with a unit vector, to project out the corresponding component of the vector equation. When a vector equation holds, all of its component equations must hold simultaneously.

In the xy axes for which x is horizontal and y is vertical, $\mathbf{F} \cdot \hat{\mathbf{i}} = m\mathbf{a} \cdot \hat{\mathbf{i}}$ gives

$$f \cos\theta - N \sin\theta = ma^x. \tag{1.29}$$

By dotting $\mathbf{F} = m\mathbf{a}$ with $\hat{\mathbf{j}}$, there results the y component,

$$N \cos\theta + f \sin\theta - mg = ma^y. \tag{1.30}$$

But when we use the $x'y'$ system of Fig. 1.1, the x'-component is obtained by dotting $\mathbf{F} = m\mathbf{a}$ with $\hat{\mathbf{i}}'$, which yields

$$f - mg \sin\theta = ma^{x'}, \tag{1.31}$$

and a similar procedure with $\hat{\mathbf{j}}'$ yields the y' component:

$$N - mg \cos\theta = 0. \tag{1.32}$$

Both pairs of (x, y) and (x', y') equations describe the same physics. We know both pairs to be equivalent because we started with the coordinate-independent relation of Eq. (1.28) and merely projected it onto these different systems of axes. However, suppose you were given the already-projected equations in the $x'y'$ system, and I was given the equations in the xy system. How could we confirm that both sets of equations describe the same net force and the same acceleration? If you could transform your $x'y'$ equations into my xy equations, or if I could transform my xy equations into your $x'y'$ equations, we would demonstrate that both sets of equations describe the same relationship. To carry out this program, we need the transformation that relates one set of coordinates to the other set.

In our example, the $x'y'$ axes can be generated from the xy axes by a rotation about the z-axis through the angle θ. With some trigonometry we construct the transformation from xyz to $x'y'z'$ and obtain

$$x' = x\cos\theta + y\sin\theta \tag{1.33}$$

$$y' = -x\sin\theta + y\cos\theta \tag{1.34}$$

$$z' = z. \tag{1.35}$$

Notice that, even though we write $z' = z$, z is not a scalar; it was never transformed in the first place, because the two coordinate systems share the same z-axis. In this instance $z' = z$ offers an example of an "identity transformation."

The transformation we have just described forms an example of an "orthogonal transformation," so called because the mutual perpendicularity of each set of axes (and their basis vectors)—their orthogonality—is preserved in rotating from (x, y, z) to (x', y', z').

By differentiating Eqs. (1.33)–(1.35) with θ fixed, we obtain

$$dx' = dx\cos\theta + dy\sin\theta \tag{1.36}$$

$$dy' = -dx\sin\theta + dy\cos\theta \tag{1.37}$$

$$dz' = dz. \tag{1.38}$$

The distance between two points does not depend on the choice of coordinate axes, because it does not even depend on the *existence* of any axes. This invariance of distance was assumed in deriving Eqs. (1.33)–(1.35) from the geometry. Distance between a given pair of points is an invariant, a scalar under orthorgonal transformations. Therefore, the transformation of Eqs. (1.36)–(1.38) must leave invariant the square of infinitesimal distance,

$$(ds')^2 = (ds)^2, \tag{1.39}$$

which can be readily confirmed:

$$\begin{aligned} (ds')^2 &= (dx')^2 + (dy')^2 + (dz')^2 \\ &= (dx\cos\theta + dy\sin\theta)^2 + (-dx\sin\theta + dy\cos\theta)^2 + (dz)^2 \\ &= (dx)^2 + (dy)^2 + (dz)^2 \\ &= (ds)^2. \end{aligned}$$

The invariance of any finite length from point a to point b therefore follows by integration over ds:

$$\Delta s = \int_a^b ds = \Delta s'. \tag{1.40}$$

Conversely, had we started from the requirement that length be an invariant, we would have been led to the orthogonal transformation (see Ex. 1.8).

In Newtonian mechanics, time intervals are explicitly assumed to be invariant, $dt' = dt$. But in an orthogonal transformation, the issue is not one of relative motion or time; observers using the primed axes measure time with the same clocks used by observers in the unprimed frame. So upon multiplying Eqs. (1.36)–(1.38) by the invariant $1/dt' = 1/dt$, the components of a moving particle's velocity vector transform identically to the coordinate displacements under a rotation of axes:

$$v^{x'} = v^x \cos\theta + v^y \sin\theta \tag{1.41}$$

$$v^{y'} = -v^x \sin\theta + v^y \cos\theta \tag{1.42}$$

$$v^{z'} = v^z. \tag{1.43}$$

Differentiating with respect to the time again, the acceleration components also transform the same as the original coordinate displacements

$$a^{x'} = a^x \cos\theta + a^y \sin\theta \tag{1.44}$$

$$a^{y'} = -a^x \sin\theta + a^y \cos\theta \tag{1.45}$$

$$a^{z'} = a^z. \tag{1.46}$$

Now we can verify a main point in the relativity of vectors (see Ex. 1.1): Although $a^{x'} \neq a^x$ and $F^{x'} \neq F^x$, the *relation* between force and acceleration is the same in both frames, $F^x = ma^x$ and $F^{x'} = ma^{x'}$. The relation $F^k = ma^k$ is said to be "covariant" between the frames, regardless of whatever system of coordinates happens to be labeled with the x^k. For this to happen, the force and acceleration vector components—indeed, the components of *any* vector—have to transform from one coordinate system to another *according to the same transfomation rule as the coordinate displacements.* This is the definition of a vector *in the context of coordinate systems.* (Note that the mass is taken to be a scalar—see Ch. 3.)

Vectors as Ratios, Displacement in Numerator

Leaving aside the gradient, think of all the vectors you met in introductory physics–velocity, momentum, acceleration, and so on. Their definitions can all be traced back to the displacement vector $d\mathbf{r}$, through the operations of scalar multiplication and parallelogram addition. For instance, velocity is $\mathbf{v} = d\mathbf{r}/dt$, momentum is $\mathbf{p} = m\mathbf{v}$, and $d\mathbf{p}/dt$ is numerically equal to the net force $\mathbf{F}$. The electric and gravitational field vectors are proportional to forces on charged and massive particles. The magnetic field, angular momentum, and Poynting's vectors are formed from other vectors through the vector product, and so on.

If the vector involves a ratio (as it does for velocity), such vectors carry the coordinate displacement in the numerator. I suggested leaving the gradient out of this consideration because, as a ratio, it has a coordinate displacement in the denominator. We will return to this point again in Chapters 3 and 7, as it lies at the heart of the distinction between a vector and its "dual."

Now we can summarize the coordinate system definition of any vector that is derived from a rescaled displacement: the quantity A^k is the component of a vector if and only if, under a coordinate transformation, it transforms the same as the corresponding coordinate displacement. For the orthogonal transformation of Eqs. (1.36)–(1.38), this means

$$A^{x'} = A^x \cos\theta + A^y \sin\theta \tag{1.47}$$

$$A^{y'} = -A^x \sin\theta + A^y \cos\theta \tag{1.48}$$

$$A^{z'} = A^z. \tag{1.49}$$

To be a robust definition of vectors that holds in generalized coordinate systems, we will have to show that such a statement applies to all kinds of coordinate transformations, not just orthogonal ones.

It may prove instructive to consider the transformation of the basis vectors. Let us start with rectangular Cartesian coordinates in the two-dimensional plane, with unit basis vectors $\hat{\mathbf{i}}$ and $\hat{\mathbf{j}}$. Under a rotation of axes, by virtue of Eqs. (1.47) and (1.48),

$$\hat{\mathbf{i}}' = \hat{\mathbf{i}}\cos\theta + \hat{\mathbf{j}}\sin\theta \tag{1.50}$$

$$\hat{\mathbf{j}}' = -\hat{\mathbf{i}}\sin\theta + \hat{\mathbf{j}}\cos\theta. \tag{1.51}$$

On the other hand, consider a transformation from rectangular to cylindrical coordinates, when both systems share a common origin. In the cylindrical coordinates ρ denotes radial distance from the origin and θ the angle from the original positive x-axis (see Appendix A). The transformation equations from (x, y) to (ρ, θ) are

$$\begin{aligned} x &= \rho\cos\theta \\ y &= \rho\sin\theta. \end{aligned}$$

A displacement in this space can be written, starting from Cartesian coordinates and converting to cylindrical ones:

$$\begin{aligned} d\mathbf{r} &= (dx)\hat{\mathbf{i}} + (dy)\hat{\mathbf{j}} \\ &= [d\rho\cos\theta - \rho\sin\theta d\theta]\hat{\mathbf{i}} + [d\rho\sin\theta + \rho\cos\theta d\theta]\hat{\mathbf{j}} \\ &= [(\cos\theta)\hat{\mathbf{i}} + (\sin\theta)\hat{\mathbf{j}}]d\rho + [(-\rho\sin\theta)\hat{\mathbf{i}} + (\rho\cos\theta)\hat{\mathbf{j}}]d\theta. \end{aligned}$$

How do we find the cylindrical basis vectors (not necessarily normalized to unit length) from this? We can see from the expression for $d\mathbf{r}$ in Cartesian coordinates that $\hat{\mathbf{i}} = \partial\mathbf{r}/\partial x$ and $\hat{\mathbf{j}} = \partial\mathbf{r}/\partial y$. Accordingly, we define

$$\vec{e}_\rho \equiv \frac{\partial\mathbf{r}}{\partial\rho} \tag{1.52}$$

and

$$\vec{e}_\theta \equiv \frac{\partial \mathbf{r}}{\partial \theta}. \tag{1.53}$$

In this way we obtain

$$\begin{aligned} \vec{e}_\rho &= (\cos\theta)\hat{\mathbf{i}} + (\sin\theta)\hat{\mathbf{j}}, \\ \vec{e}_\theta &= \rho(-\sin\theta)\hat{\mathbf{i}} + \rho(\cos\theta)\hat{\mathbf{j}}. \end{aligned}$$

Notice that Eqs. (1.47)-(1.49) describe the *same* vector in terms of how its new *components* are superpositions of the old ones, whereas Eqs. (1.50) and (1.51) describe *new* basis *vectors* as superpositions of the old ones.

The g_{nm} coefficients, computed according to Eq. (1.18), $g_{nm} = \vec{e}_n \cdot \vec{e}_m$, are $g_{\rho\rho} = 1$, $g_{\rho\theta} = 0$, $g_{\theta\rho} = 0$, and $g_{\theta\theta} = \rho^2$. Thus an infinitesimal distance between two points, in cylindrical coordinates, becomes

$$d\mathbf{r} \cdot d\mathbf{r} = (d\rho)^2 + \rho^2 (d\theta)^2. \tag{1.54}$$

To obtain the unit basis vectors $\hat{\boldsymbol{\rho}}$ and $\hat{\boldsymbol{\theta}}$ familiar in cylindrical coordinates, divide the basis vectors, as found from the transformation, by their magnitudes and obtain

$$\hat{\boldsymbol{\rho}} = \frac{\vec{e}_\rho}{|\vec{e}_\rho|} = \frac{\vec{e}_\rho}{1} = \vec{e}_\rho \tag{1.55}$$

and

$$\hat{\boldsymbol{\theta}} = \frac{\vec{e}_\theta}{|\vec{e}_\theta|} = \frac{\vec{e}_\theta}{\rho}, \tag{1.56}$$

or $\vec{e}_\rho = \hat{\boldsymbol{\rho}}$ but $\vec{e}_\theta = \rho\hat{\boldsymbol{\theta}}$. This distinction between basis vectors and unit basis vectors accounts for some of the factors such as $1/\rho$ that appears in expressions for the gradient in cylindrical coordinates, and $1/(r\sin\theta)$ in spherical coordinates.

Vectors as Ratios, Displacement in Denominator

As alluded to above, we will meet another kind of "vector" whose prototype is the gradient, which takes derivatives of a scalar function with respect to coordinate displacements. For the gradient I put "vector" in quotation marks because in the gradient the coordinate displacement appears "downstairs," as the "denominator" of a derivative, such as $\frac{\partial}{\partial x}$ or $\frac{\partial}{\partial \theta}$. In contrast, the vectors we discussed prior to the gradient were proportional to rescaled coordinate displacements, with components such as $\frac{dx}{dt}$ or $\frac{d\theta}{dt}$. Following tradition, we use superscripts to label vector components that are proportional to "numerator displacements," such as

$$v^x = \frac{dx}{dt}, \tag{1.57}$$

and we will use subscripts for "denominator displacements" such as

$$\nabla_x U = \frac{\partial U}{\partial x}. \tag{1.58}$$

In the classic texts on tensor calculus "coordinate-displacement-in-numerator" vectors are called "contravariant" vectors (with superscripts) and the "coordinate-displacement-in-denominator" vectors are called "covariant" vectors (with subscripts). Unfortunately, in tensor calculus the word "covariant" carries three distinct, context-dependent meanings: (1) One usage refers to the covariance of an *equation*, which transforms under a coordinate transformation the same as the coordinates themselves. That was the concept motivating the inclined plane example of this section, illustrating how the relationship expressed in the equation transcends the choice of a coordinate system. (2) The second usage, just mentioned, makes the distinction between "contravariant" and "covariant" *vectors*. Alternative terms for "covariant vector" speak instead of "dual vectors" (Ch. 7) or "1-forms" (Ch. 8). (3) The third usage for the word "covariant" arises in the context of *derivatives* of tensors, where the usual definition of derivative must be extended to the so-called covariant derivative, which is sometimes necessary to guarantee the derivative of a tensor being another tensor (see Ch. 4). Thus, "covariant derivatives" (usage 3) are introduced so equations with derivatives of tensors will "tranform covariantly" (usage 1). Of course, among these will be the derivatives of "covariant vectors" (usage 2)!

I emphasize that, for now and throughout Chapter 2, usages (2) and (3) do not yet concern us, because we will still be working exclusively in Euclidean spaces where contravariant and covariant vectors are identical and covariant derivatives are unnecessary. But to develop good habits of mind for later developments, we use superscripts for vector components from the start.

1.7 Transformation Coefficients as Partial Derivatives

Let us take another look at the formal tensor definition presented at the beginning of this book. Where do its partial derivatives come from? In particular, go back to Eq. (1.33) which says

$$x' = x\cos\theta + y\sin\theta. \tag{1.59}$$

Notice that x' in this instance is a function of x and y (with the rotation angle θ considered a fixed parameter). Then

$$\frac{\partial x'}{\partial x} = \cos\theta \tag{1.60}$$

and

$$\frac{\partial x'}{\partial y} = \sin\theta, \tag{1.61}$$

and similar expressions may be written for the other transformed coordinates (viz., $\partial x'/\partial z = 0$). Now Eq. (1.36) can be written

$$dx' = \frac{\partial x'}{\partial x}dx + \frac{\partial x'}{\partial y}dy + \frac{\partial x'}{\partial z}dz, \tag{1.62}$$

and likewise for dy' and dz'.

This result illustrates what would have been obtained in general by thinking of each new coordinate as a function of all the old ones, $x' = x'(x, y, z)$, $y' = y'(x, y, z)$, and $z' = z'(x, y, z)$. Evaluating their differentials using the chain rule of multivariable calculus, we would write

$$dx' = \frac{\partial x'}{\partial x}dx + \frac{\partial x'}{\partial y}dy + \frac{\partial x'}{\partial z}dz, \tag{1.63}$$

$$dy' = \frac{\partial y'}{\partial x}dx + \frac{\partial y'}{\partial y}dy + \frac{\partial y'}{\partial z}dz, \tag{1.64}$$

$$dz' = \frac{\partial z'}{\partial x}dx + \frac{\partial z'}{\partial y}dy + \frac{\partial z'}{\partial z}dz. \tag{1.65}$$

Such equations hold not only for orthogonal transformations, but for *any* transformation for which the partial derivatives exist.

For example, in switching from rectangular to spherical coordinates, the radial coordinate r is related to (x, y, z) according to

$$r = \sqrt{x^2 + y^2 + z^2}. \tag{1.66}$$

Therefore, under a displacement (dx, dy, dz) the corresponding change in the r-coordinate is

$$dr = \frac{\partial r}{\partial x}dx + \frac{\partial r}{\partial y}dy + \frac{\partial r}{\partial z}dz, \tag{1.67}$$

where

$$\frac{\partial r}{\partial x} = \frac{x}{\sqrt{x^2 + y^2 + z^2}} \tag{1.68}$$

and so on.

The components of any vector made from a displacement in the numerator must transform from one coordinate system to another *the same as the coordinate differentials.* Whenever the new coordinates x'^i are functions of the old ones, the x^j, the displacements transform as

$$dx'^i = \sum_j \frac{\partial x'^i}{\partial x^j} dx^j. \tag{1.69}$$

Therefore, vectors are, by definition, quantities whose components transform according to

$$A'^i = \sum_j \frac{\partial x'^i}{\partial x^j} A^j, \tag{1.70}$$

a statement on the relativity of vectors!

1.8 What Is a Theory of Relativity?

At the beginning of this book we wondered why tensor components are traditionally defined not in terms of what they *are*, but in terms of *how they change* under a coordinate transformation. Such focus on changes in coordinates means that tensors are important players in theories of relativity.

What is a theory of relativity? Any theory of relativity is a set of principles and procedures that tell us what happens to the numerical values of quantities, and to relationships between them, when we switch from one member of a suitably defined class of coordinate systems to another. Consider a couple of thought experiments.

For the first thought experiment, imagine a particle of electric charge q sitting at rest in the lab frame. Let it be located at the position identified by the vector $\mathbf{s}$ relative to the origin. This charge produces an electrostatic field described by a vector $\mathbf{E}$. At the location specified by the position vector $\mathbf{r}$ in the lab frame coordinate system, $\mathbf{E}$ is given by Coulomb's law,

$$\mathbf{E}(\mathbf{r}) = kq\frac{(\mathbf{r}-\mathbf{s})}{|\mathbf{r}-\mathbf{s}|^3}, \tag{1.71}$$

where $k = \frac{1}{4\pi\epsilon_o}$ is Coulomb's constant. Since this charge is at rest in the lab frame, its magnetic field there vanishes, $\mathbf{B} = \mathbf{0}$.

A rocket moves with constant velocity $\mathbf{v}_r$ through the lab frame. The subscript "r" denotes relative velocity between the two frames. In the coasting rocket frame, an observer sees the charge q zoom by with velocity $-\mathbf{v}_r$.

A theory of relativity tackles such problems as this: given $\mathbf{E}$ and $\mathbf{B}$ in the lab frame, and given the relative velocity between the lab and the coasting rocket frames, what is the electric field $\mathbf{E}'$ and the magnetic field $\mathbf{B}'$ observed in the rocket frame? We seek the new fields in terms of the old ones, and in terms of parameters that relate the two frames:

$$\mathbf{E}' = \mathbf{E}'(\mathbf{E},\mathbf{B},\mathbf{v}_r) \tag{1.72}$$

and

$$\mathbf{B}' = \mathbf{B}'(\mathbf{E},\mathbf{B},\mathbf{v}_r). \tag{1.73}$$

It was such problems that led Albert Einstein to develop the special theory of relativity. His celebrated 1905 paper on this subject was entitled "On the Electrodynamics of Moving Bodies." He showed that, for two inertial frames (where $\mathbf{v}_r = const.$),

$$\mathbf{E}' = \gamma_r\left[\mathbf{E} + (\frac{\mathbf{v}_r}{c}\times\mathbf{B})\right] - \frac{\gamma_r^2}{\gamma_r+1}\frac{\mathbf{v_r}}{c}\left(\frac{\mathbf{v_r}}{c}\cdot\mathbf{E}\right) \tag{1.74}$$

and

$$\mathbf{B}' = \gamma_r\left[\mathbf{B} - (\frac{\mathbf{v}_r}{c}\times\mathbf{E})\right] - \frac{\gamma_r^2}{\gamma_r+1}\frac{\mathbf{v_r}}{c}\left(\frac{\mathbf{v_r}}{c}\cdot\mathbf{B}\right), \tag{1.75}$$

where $\gamma_r \equiv [1 - (\frac{\mathbf{v}_r}{c})^2]^{-\frac{1}{2}}$ and c denotes the speed of light in vacuum, which is postulated in special relativity to be invariant among all inertial reference frames.

For the second thought experiment, imagine a positron moving through the lab and colliding with an electron. Suppose in the lab frame the target electron sits at rest and the incoming positron carries kinetic energy K and momentum $\mathbf{p}$. The electron-positron pair annihilates into electromagnetic radiation. Our problem is to predict the minimum number of photons that emerge from the collision, along with their energies and momenta as measured in the lab frame. Working in the lab frame, denoting a free particle's energy as E, we write the simultaneous equations expressing conservation of energy and momentum in the positron-electron collision that produces N photons. For the energy,

$$K + 2mc^2 = E_1 + E_2 + E_3 + ... + E_N, \tag{1.76}$$

and for momentum,

$$\mathbf{p} = \mathbf{p}_1 + \mathbf{p}_2 + ...\mathbf{p}_N. \tag{1.77}$$

From special relativity we also know that, outside the interaction region where each particle moves freely, its energy, momentum, and mass are related by.

$$E^2 - (pc)^2 = (mc^2)^2, \tag{1.78}$$

where $E = mc^2\gamma = K + mc^2$, $\mathbf{p} = m\mathbf{v}\gamma$, and $\gamma \equiv [1 - (v/c)^2]^{-\frac{1}{2}}$ (this $\mathbf{v}$ and its γ are measured for velocities *within* a given frame; they are not to be confused with $\mathbf{v}_r$ and γ_r that hold for relative velocity *between* frames). For a photon, $E = pc$ since it carries zero mass.

Our task is to determine the minimum value of N. In the lab frame, the total energy and momentum of the system are nonzero, so their conservation requires $N > 0$. Could N be as small as 1? A second perspective may be helpful.

In the center-of-mass frame of the positron-electron system, before the collision the electron and positron approach each other with equal and opposite momentum. By the conservation of momentum, the momentum of all the photons after the collision must therefore sum to zero. We immediately see that there can be no less than two photons after the collision, and in that case their momenta in the center-of-mass frame must be equal and opposite. We also know that their energies are equal in this frame, from $E' = p'c$. To find each photon's momentum and energy back in the lab frame, it remains only to transform these quantities from the center-of-mass frame back into the lab frame. Now we are *using* a theory of relativity effectively.

For this strategy to work, the principles on which the solution depends must apply within *both* frames. If momentum is conserved in the lab frame, then momentum must be conserved in the center-of-mass frame, even though the system's total momentum is nonzero in the former and equals zero in the latter. But whatever its numerical value, the statement "total momentum before the collision equals total momentum after the collision" holds within *all* frames, regardless of the coordinate system.

Physicists have to gain expertise in several theories of relativity, including:

* Changing from one kind of coordinate grid to another (rectangular, spherical, cylindrical, and so forth).

* Rotations and translations of coordinate axes.

* Relative motion, or "boosts" between inertial reference frames. These include Newtonian relativity, based on the postulates of absolute space and absolute time, which find expression in the Galilean transformation. Boosts between inertial frames are also the subject of special relativity, expressed in the Lorentz transformation, based on the postulate of the invariance of the speed of light.

* Changing from an inertial frame to an accelerated frame.

Other transformations of coordinates, such as conformal mappings, rescalings, space inversions, and time reversals, have their uses. But the theories of relativity mentioned above suffice to illustrate an essential point: any theory of relativity is built on the foundation of *what stays the same* when changing coordinate systems. A theory of relativity is founded on its scalars.

The choice of a reference frame is a matter of convenience, not of principle. Any "laws of nature" that we propose must transcend the choice of this or that coordinate system. Results derived in one frame must be translatable into any other frame. To change reference frames is to carry out a coordinate transformation.

Any equation written in tensors is independent of the choice of coordinates, in the sense that tensors transform consistently and unambiguously from one reference frame to another—as go the coordinates, so go the tensors. This relativity principle is articulated by saying that our laws of physics must be written "covariantly," written as tensors.

For example, in Newtonian mechanics, the coordinate-free expressions

$$\mathbf{F} = m\frac{d^2\mathbf{r}}{dt^2} \tag{1.79}$$

can be expressed covariantly in terms of generalized coordinates x^i by writing

$$F^i = m\frac{d^2x^i}{dt^2}. \tag{1.80}$$

This expression is reference frame independent because it can be transformed, as easily as the coordinates themselves, between any two coordinate systems related by a well-defined transformation. This follows from simply multiplying $F^i = ma^i$ with transformation coefficients:

$$\frac{\partial x'^j}{\partial x^i}F^i = m\frac{\partial x'^j}{\partial x^i}a^i. \tag{1.81}$$

Summing over i, we recognize the left-hand side as F'^j and the right-hand side as ma'^j and have thus moved smoothly from $F^i = ma^i$ to $F'^j = ma'^j$. In that sense Eq. (1.80) is just as coordinate frame independent as $\mathbf{F} = m\mathbf{a}$.

Finally, in this section we make an important technical note, which holds for any transformation, not just the orthogonal ones we have been using for illustrations. Under a change of coordinates from the unprimed to the primed system we have seen that

$$dx'^i = \sum_j \frac{\partial x'^i}{\partial x^j} dx^j. \tag{1.82}$$

Likewise, for the reverse transformation from the primed to the unprimed system we would write

$$dx^j = \sum_n \frac{\partial x^j}{\partial x'^n} dx'^n. \tag{1.83}$$

By putting Eq. (1.83) into Eq. (1.82), which reverses the original transformation and returns to what we started with, slogging through the details we see that

$$dx'^i = \sum_j \sum_n \frac{\partial x'^i}{\partial x^j} \frac{\partial x^j}{\partial x'^n} dx'^n. \tag{1.84}$$

The double sum must pick out dx'^i and eliminate all the other dx'^n displacements; in other words, this requires

$$\sum_j \frac{\partial x'^i}{\partial x^j} \frac{\partial x^j}{\partial x'^n} = \delta^{in}, \tag{1.85}$$

where δ^{in} is the Kronecker delta.

1.9 Vectors Represented as Matrices

Any vector in three-dimensional space that has been projected onto a coordinate system may be written as the ordered triplet, $\mathbf{A} = (a, b, c)$. The components of the vector can also be arranged into a column matrix denoted $|A\rangle$:

$$|A\rangle \equiv \begin{bmatrix} a \\ b \\ c \end{bmatrix}. \tag{1.86}$$

The "arrow" $\mathbf{A}$, the ordered triplet (a, b, c), and the matrix $|A\rangle$ carry the same information.

This "bracket notation" was introduced into physics by Paul A. M. Dirac, when he developed it for the abstract "state vectors" in the infinite-dimensional, complex-number-valued "Hilbert space" of quantum mechanics. Dirac's vector notation can also be used to describe vectors in three-dimensional Euclidean space and in the four-dimensional spacetimes of special and general relativity.

In the notation of Dirac brackets, a vector equation such as $\mathbf{F} = m\mathbf{a}$ may be expressed as the equivalent matrix equation

$$|F\rangle = m|a\rangle. \tag{1.87}$$

(A notational distinction between $|\mathbf{F}\rangle$ and $|F\rangle$ is not necessary; in this book $\mathbf{F}$ denotes an "arrow" and $|F\rangle$ denotes the corresponding matrix, but they carry identical information.) To write the scalar product in the language of matrix multiplication, corresponding to a vector $|A\rangle$, we need to introduce its transpose or row matrix $\langle A|$:

$$\langle A| = \begin{bmatrix} a & b & c \end{bmatrix}. \tag{1.88}$$

(For vectors whose components may be complex numbers, as in quantum mechanics, the elements of $\langle A|$ are also the complex conjugates of those in $|A\rangle$. The transpose and complex conjugate of a matrix is said to be the "adjoint" of the original matrix. When all the matrix elements are real numbers, the adjoint is merely the transpose.) Let another vector $|B\rangle$ be

$$|B\rangle \equiv \begin{bmatrix} e \\ f \\ g \end{bmatrix}. \tag{1.89}$$

According to the rules of matrix multiplication, the scalar product $\mathbf{A} \cdot \mathbf{B}$ will be evaluated, in matrix language, as "row times column," $\langle A|B\rangle$. Upon multiplying everything out, it yields

$$\langle A|B\rangle = ae + bf + cg, \tag{1.90}$$

as expected. The scalar product is also called the "contraction" or the "inner product" of the two vectors, because all the components get reduced to a single number. That number will be shown to be a scalar, an invariant among the family of coordinate systems that can be transformed into one another.

An "outer product" $|A\rangle\langle B|$ can also be defined, which takes two three-number quantities (for vectors in three dimensions) and, by the rules of matrix multiplication, makes nine numbers out of them, represented by a square matrix:

$$\begin{aligned} |A\rangle\langle B| &= \begin{bmatrix} a \\ b \\ c \end{bmatrix} \begin{bmatrix} e & f & g \end{bmatrix} \\ &= \begin{bmatrix} ae & af & ag \\ be & bf & bg \\ ce & cf & cg \end{bmatrix}. \end{aligned}$$

The familiar unit basis vectors for xyz axes can be given the matrix representations

$$\hat{\mathbf{i}} \equiv |1\rangle \equiv \begin{bmatrix} 1 \\ 0 \\ 0 \end{bmatrix}, \tag{1.91}$$

$$\hat{\mathbf{j}} \equiv |2\rangle \equiv \begin{bmatrix} 0 \\ 1 \\ 0 \end{bmatrix}, \tag{1.92}$$

and

$$\hat{\mathbf{k}} \equiv |3\rangle \equiv \begin{bmatrix} 0 \\ 0 \\ 1 \end{bmatrix}. \tag{1.93}$$

The inner product of any two of these orthonormal unit basis vectors results in the Kronecker delta,

$$\langle i|j\rangle = \delta^{ij}. \tag{1.94}$$

The sum of their outer products gives the "completeness relation"

$$|1\rangle\langle 1| + |2\rangle\langle 2| + |3\rangle\langle 3| = \mathbf{1}, \tag{1.95}$$

where $\mathbf{1}$ means the 3×3 unit matrix, whose elements are the δ^{ij}. Any set of orthonomal unit vectors that satisfy the completeness relation form a "basis"; they are linearly independent and span the space.

As a superposition of weighted unit vectors, any vector $|A\rangle$ can be written in bracket notation as

$$|A\rangle = a|1\rangle + b|2\rangle + c|3\rangle \tag{1.96}$$

and similarly for $\langle B|$,

$$\langle B| = e\langle 1| + f\langle 2| + g\langle 3|. \tag{1.97}$$

Consider the outer product $\mathbf{M} \equiv |A\rangle\langle B|$. Its element standing in the ith row and jth column, M^{ij}, is the same as

$$M^{ij} = \langle i|\mathbf{M}|j\rangle, \tag{1.98}$$

as you can readily show. For instance, by the rules of matrix multiplication, in our previous example $M^{12} = \langle 1|A\rangle\langle B|2\rangle = af$.

As we saw earlier in other notation, through basis vectors the algebra of vectors exhibits a synthesis/analysis dichotomy. On the "synthesis" side, any vector in the space can be synthesized by rescaling the unit basis vectors with coefficients (the "components"), and then summing these to give the complete vector. On the "analysis" side, a given vector can be resolved into its components. The Dirac bracket notation, along with creative uses of the unit matrix $\mathbf{1}$, allow the synthesis/analysis inverse operations and coordinate transformations to be seen together with elegance. Begin with the completeness relation for some set of orthonormal basis vectors,

$$\mathbf{1} = \sum_{\alpha} |\alpha\rangle\langle\alpha|. \tag{1.99}$$

Multiply the completeness relation from the right by $|A\rangle$ to obtain

$$\begin{aligned} |A\rangle &= \sum_{\alpha} |\alpha\rangle\langle\alpha|A\rangle \\ &= \sum_{\alpha} A^{\alpha}|\alpha\rangle, \end{aligned}$$

illustrating how, if the A^{α} are given, we can synthisize the entire vector $|A\rangle$ by a superposition of weighted unit vectors. Conversely, the analysis equation has presented itself to us as

$$A^{\alpha} \equiv \langle\alpha|A\rangle, \tag{1.100}$$

which shows how we can find each of its components if the vector is given.

Let us examine vector transformations in the language of matrices. The properties of unit basis vectors themselves can be used to change the representation of a vector $|A\rangle$ from an old basis (denoted here with Latin letters) to a new basis (denoted here with Greek letters instead of primes), provided that each basis set respects the completeness relation, where

$$\sum_{i} |i\rangle\langle i| = \mathbf{1} \tag{1.101}$$

in the old basis and

$$\sum_{\alpha} |\alpha\rangle\langle\alpha| = \mathbf{1} \tag{1.102}$$

in the new basis. Begin by multiplying $|A\rangle$ with $\mathbf{1}$, as follows:

$$\begin{aligned} |A\rangle &= \mathbf{1}|A\rangle \\ &= \sum_{i} |i\rangle\langle i|A\rangle \\ &= \sum_{i} |i\rangle A^{i}. \end{aligned}$$

The transformation equations, in bracket notation, can be derived by multiplying this superposition over the $|i\rangle$ basis by $\langle\alpha|$ to obtain

$$\langle\alpha|A\rangle = \sum_{i} \langle\alpha|i\rangle A^{i}, \tag{1.103}$$

which is a Dirac notation version of our familiar vector transformation rule (restoring primes),

$$A^{\alpha'} = \sum_{i} \frac{\partial x'^{\alpha}}{\partial x^{i}} A^{i}. \tag{1.104}$$

Comparing these results yields

$$\frac{\partial x'^{\alpha}}{\partial x^{i}} = \langle\alpha|i\rangle \equiv \Lambda_{\alpha i}, \tag{1.105}$$

where Λ denotes the matrix of transformation coefficients, with $\Lambda_{\alpha i}$ the matrix element in the αth row and jth column (the Λ matrix elements are not tensor components themselves; subscripts are chosen so the notation will be consistent with tensor equations when distinctions eventually must be made between upper and lower indices). The component-by-component expressions for the transformation of a vector can be gathered into one matrix equation for turning one column vector into another,

$$|A'\rangle = \Lambda|A\rangle. \tag{1.106}$$

The row vector $\langle A'|$ is defined, according to the rules of matrix multiplication, as

$$\langle A'| = \langle A|\Lambda^\dagger, \tag{1.107}$$

where $\Lambda^\dagger$ denotes the adjoint of the square matrix Λ, the transpose and complex conjugate of the original matrix. Notice a very important point: To preserve the magnitude of a vector as an invariant, we require that

$$\langle A'|A'\rangle = \langle A|A\rangle. \tag{1.108}$$

But in terms of the transformation matrices, $\langle A'|A'\rangle$ means

$$\langle A'|A'\rangle = \langle A|\Lambda^\dagger\Lambda|A\rangle\ . \tag{1.109}$$

Therefore, the Λ-matrices must be "unitary,"

$$\Lambda^\dagger\Lambda = \mathbf{1} \tag{1.110}$$

which means that the adjoint equals the multiplicative inverse,

$$\Lambda^\dagger = \Lambda^{-1}. \tag{1.111}$$

Term by term, these relations say that, with

$$\Lambda_{\alpha i} = \frac{\partial x'^\alpha}{\partial x^i}, \tag{1.112}$$

then

$$\begin{aligned}(\Lambda^\dagger\Lambda)_{jk} &= \sum_\alpha (\Lambda^\dagger)_{j\alpha}\Lambda_{\alpha k} \\ &= \sum_\alpha \frac{\partial x^j}{\partial x'^\alpha}\frac{\partial x'^\alpha}{\partial x^k} \\ &= \delta^{jk}.\end{aligned}$$

The adjoint of a transformation matrix being equal to its multiplicative inverse is the signature of a "unitary transformation." For a rotation of axes, all the matrix elements are real, so the adjoint of the transformation matrix is merely the transpose; therefore, the transpose is the multiplicative inverse in an "orthogonal transformation."

We have said that $\langle A|$ is the "dual" of $|A\rangle$. Throughout this subject, especially beginning in Chapter 3, the concept of "duality" will be repeatedly enountered. In general, any mathematical object $\mathcal{M}$ needs corresponding to it a dual $\mathcal{M}^*$ whenever $\mathcal{M}$, by itself, cannot give a single real number as the measure of its magnitude. For instance, consider the complex number $z = x + iy$, where x and y are real numbers and $i^2 = -1$. To find the distance $|z|$ between the origin and a point in the complex plane requires not only z itself but also its complex conjugate $z^* = x - iy$. For then the distance is $|z| = \sqrt{z^* z} = \sqrt{x^2 + y^2}$. The square of z does not give a distance because it is a complex number, $z^2 = x^2 - y^2 + 2ixy$. To define length in the complex plane, z needs its dual, z^*.

Similarly, in quantum mechanics we deal with the wave function $\psi(x)$, and the square of ψ is supposed to give us the probability density of locating a particle in the neigborhood of x. However, ψ is a complex function, with both a real and an imaginary part. Thus, by the "square" of ψ we mean $\psi^*\psi$, where $*$ means complex conujugate. Thus, ψ^* is the dual of ψ beause both the wave function and its dual are necessary to measure its "magnitude," in this case the probability of a particle being found between $x = a$ and $x = b$ as a kind of inner product,

$$\int_a^b \psi^* \psi dx. \tag{1.113}$$

Probabilities must be real numbers on the interval [0,1].

In general, if $\mathcal{M}$ is a mathematical object and some measure of its magnitude or length is required, then one defines a corresponding dual $\mathcal{M}^*$ defined such that the length extracted from $\mathcal{M}^*\mathcal{M}$ is a real number.

In the next chapter we will go beyond scalars and vectors and make the acquaintance of tensors with two or more indices as they arise in the context of physics applications.

1.10 Discussion Questions and Exercises

Discussion Questions

Q1.1 Write a "diary entry" describing your first encounter with tensors. How were you introduced to them? What did you think of tensors on that occasion?

Q1.2 In introductory physics a vector is described as "a quantity having magnitude and direction," which respects scalar multiplication and parallelogram addition. Discuss the consistency between this definition and the definition of a vector in terms of its behavior under coordinate transformations.

Q1.3 List several of the vectors (not including the gradient) that you have met in physics. Trace each one back to a rescaled displacement vector; in other words, show how a vector's "genealogy" ultimately traces back to the displacement vector $d\mathbf{r}$, through scalar multiplication and parallelogram addition.

Q1.4 Critique this equation:

$$\hat{\mathbf{k}} = \alpha\hat{\mathbf{i}} + \beta\hat{\mathbf{j}}. \tag{1.114}$$

Q1.5 What are the advantages of using basis vectors normalized to be dimensionless, of unit length, and orthogonal? Do basis vectors *have* to be orthogonal, of unit length, or dimensionless?

Exercises

1.1 Confirm by explicit calculation that $a^x \cos\theta + a^y \sin\theta$ gives $a^{x'}$ in Eq. (1.44), by using the a^x and a^y from Eqs. (1.29)–(1.30).

1.2 Using the Levi-Civita symbol, prove that

$$\mathbf{A} \times (\mathbf{B} \times \mathbf{C}) = (\mathbf{A} \cdot \mathbf{C})\mathbf{B} - (\mathbf{A} \cdot \mathbf{B})\mathbf{C}. \tag{1.115}$$

You may need this identity (see, e.g., Marion and Thornton, p. 45):

$$\sum_k \varepsilon^{ijk}\varepsilon^{lmk} = \delta^{il}\delta^{jm} - \delta^{im}\delta^{jl}. \tag{1.116}$$

1.3 Show that the scalar product, $\mathbf{A} \cdot \mathbf{B}$ or $\langle A|B\rangle$, is a scalar under an orthogonal transformation.

1.4 Show that the two definitions of the scalar product,

$$\mathbf{A} \cdot \mathbf{B} = |\mathbf{A}||\mathbf{B}|\cos\theta \tag{1.117}$$

and

$$\mathbf{A} \cdot \mathbf{B} = A^x B^x + A^y B^y + A^z B^z, \tag{1.118}$$

are equivalent.

1.5 Show that the two definitions of the vector product,

$$\mathbf{A} \times \mathbf{B} = (|\mathbf{A}||\mathbf{B}|\sin\theta)\hat{\mathbf{n}} \tag{1.119}$$

and

$$\mathbf{A} \times \mathbf{B} = \begin{vmatrix} \hat{\mathbf{i}} & \hat{\mathbf{j}} & \hat{\mathbf{k}} \\ A^x & A^y & A^z \\ B^x & B^y & B^z \end{vmatrix},$$

are equivalent.

1.6 (a) Write the Λ matrix for the simple orthogonal transformation that rotates the xy axes through the angle θ about the z axis to generate new $x'y'z'$ axes.
(b) Show by explicit calculation for the Λ of part (a) that $\Lambda^\dagger = \Lambda^{-1}$.
(c) Show that $|\Lambda|^2 = 1$, where $|\Lambda|$ denotes the determinant of Λ. Which option occurs here, $|\Lambda| = +1$ or $|\Lambda| = -1$?

1.7 Consider an inversion of axes (reflection of all axes through the origin, a "space inversion"), $x'^i = -x^i$.
(a) What is Λ? What is $|\Lambda|$?
(b) Under a space inversion, a vector $\mathbf{A}$ goes to $\mathbf{A}' = -\mathbf{A}$ and $\mathbf{B}$ goes to $\mathbf{B}' = -\mathbf{B}$. How does $\mathbf{A}' \cdot \mathbf{B}'$ compare to $\mathbf{A} \cdot \mathbf{B}$, and how does $\mathbf{A}' \times \mathbf{B}'$ compare to $\mathbf{A} \times \mathbf{B}$? When the distinction is necessary, vectors that change sign under a space inversion are called "polar vectors" (or simply "vectors") and vectors that preserve their sign under a space inversion are called "axial vectors" (or "pseudovectors").
(c) Show that $(\mathbf{A} \times \mathbf{B}) \cdot \mathbf{C}$ is a scalar under a rotation of axes, but that it changes sign under a space inversion. Such a quantity is called a "pseudoscalar."
(d) Is $(\mathbf{A} \times \mathbf{B}) \times (\mathbf{C} \times \mathbf{D})$ a vector or a pseudovector?

1.8 Consider a rotation of axes through angle θ about the z-axis, taking (x, y) coordinates into (x', y') coordinates. Assuming a linear transformation, parameterize it as

$$\begin{aligned} x' &= Ax + By \\ y' &= Cx + Dy. \end{aligned}$$

By requiring length to be an invariant, derive the results expected for A, B, C and D, in terms of θ.

1.9 Consider a rotation about the z-axis that carries (x, y) to (x', y'). Show that, if the rotation matrix from xyz to $x'y'z'$ (where $z = z'$) is parameterized as

$$\Lambda = \begin{vmatrix} a & b & 0 \\ c & d & 0 \\ 0 & 0 & 1 \end{vmatrix},$$

then just from the requirement that $\Lambda^\dagger = \Lambda^{-1}$, it follows that $a = d = \cos\theta$ and $b = -c = \sin\theta$. Therefore, a unitarity transformation requires the coordinate transformation to be equivalent to a rotation of axes. (Sometimes, as in the unitary transformations of quantum mechanics, the axes exist within a highly abstract space.)

Chapter 2

Two-Index Tensors

Tensors typically enter the consciousness of undergraduate physicsts when two-index tensors are first encountered. Let us meet some examples of two-index tensors as they arise in physics applications.

2.1 The Electric Susceptibility Tensor

Our first exhibit introduces a two-index tensor that describes a phenomenological effect that occurs when dielectric material gets immersed in an externally applied electric field. Place a slab of plastic or glass between the plates of a parallel-plate capacitor and then charge the capacitor. Each molecule in the dielectric material gets stretched by the electric force and acquires a new or enhanced electric dipole moment. The macroscopic polarization in the dielectric can be described with the aid of a vector **P**, the density of electric dipole moments. If there are N molecules per unit volume, and their spatially averaged electric dipole moment is $\langle \mathbf{p} \rangle$, then $\mathbf{P} = N\langle \mathbf{p} \rangle$. In most dielectrics, **P** is proportional to the **E** that produces it, and we may write

$$\mathbf{P} = \epsilon_o \chi \mathbf{E}, \tag{2.1}$$

which defines the "electric susceptibility" χ of the material (the permittivity of vacuum, ϵ_o, is customarily included in the definition of the susceptibility to make χ dimensionless). In such materials, if **E** points north, then **P** has only a northward component.

However, some dielectrics exhibit the weird property that when **E** points north, then **P** points in some other direction! In such dielectrics, the y-component of **P** depends in general not only on the y component of **E** but also on the latter's x and z components, and we must write

$$P^y = \epsilon_o(\chi^{yx}E^x + \chi^{yy}E^y + \chi^{yz}E^z). \tag{2.2}$$

Similar expressions hold for P^x and P^z. The nine coefficients χ^{ij} are called the components of the "electric susceptability tensor." These components can be calculated from quantum mechanics applied to the field-molecule interaction. Our business here is to note that the set of coefficients χ^{ij} fall readily to hand when the "response" vector $\mathbf{P}$ is rotated, as well as rescaled, relative to the "stimulus" vector $\mathbf{E}$.

Introducing numerical superscripts to denote components, where x^1 means x, x^2 means y, and x^3 means z, we may write

$$P^i = \epsilon_o \sum_{j=1}^{3} \chi^{ij} E^j, \tag{2.3}$$

where $i = 1, 2$, or 3. The nine χ^{ij} can be arranged in a 3×3 matrix. The susceptibility tensor can be shown from energy considerations to be symmetric (see Ex. 2.15),

$$\chi^{ij} = \chi^{ji}, \tag{2.4}$$

which reduces the number of independent components from 9 to 6.

Just because these susceptibility coefficients can be arranged in a matrix does not make them tensor components. For an array of numbers to be tensor components, specific conditions must be met concerning their behavior under coordinate transformations. Without resorting for now to a formal proof, what can we anticipate about transforming the χ^{ij}? The very existence of electric polarization depends on the existence of a displacement between a positive and a negative charge. The coefficients χ^{ij} should therefore be proportional to such displacements, and we know how displacements transform. This intuition is confirmed when the susceptibility tensor components are computed from quantum mechanics applied to a molecule in an electric field (see Ex. 2.17).

2.2 The Inertia Tensor

Typically one of the first two-index tensors to be confronted in one's physics career, the inertia tensor emerges in a study of rigid-body mechanics. In introductory physics one meets its precursor, the moment of inertia, when studying the dynamics of a rigid body rotating about a fixed axis. Let us review the moment of inertia and then see how it generalizes to the inertia tensor when relaxing the fixed axis restriction.

Conceptualize a rigid body as an array of infinitesimal chunks, or particles, each of mass dm. The dm located at the distance s from the fixed axis of rotation gets carried in a circle about that axis with speed $v = s\omega$, where $\boldsymbol{\omega}$ denotes the rotation's angular velocity vector. To calculate the angular momentum $\mathbf{L}$ about the axis of rotation, because angular momentum is additive, we compute the sum

$$\begin{aligned}\mathbf{L} &= \int(\mathbf{r}\times\mathbf{v})dm \\ &= \boldsymbol{\omega}\int s^2 dm.\end{aligned}$$

Introducing the "moment of inertia,"

$$I \equiv \int s^2 dm, \tag{2.5}$$

we may write

$$|L\rangle = I|\omega\rangle. \tag{2.6}$$

For instance, the moment of inertia of a uniform sphere of mass m and radius R, evaluated about an axis passing through the sphere's center, is $\frac{2}{5}mR^2$.

Significantly, for rotations about a fixed axis the angular momentum vector $\mathbf{L}$ and the angular velocity vector $\boldsymbol{\omega}$ are parallel. For instance, if the rigid body spins about the z axis, then component by component Eq. (2.6) says

$$\begin{aligned}L^x &= 0 \\ L^y &= 0 \\ L^z &= I\omega.\end{aligned}$$

The rigid body's rotational kinetic energy K also follows from energy being additive:

$$\begin{aligned}K &= \frac{1}{2}\int v^2 dm \\ &= \frac{1}{2}I\omega^2.\end{aligned}$$

We may also write K as

$$\begin{aligned}K &= \frac{1}{2}\mathbf{L}\cdot\boldsymbol{\omega} \\ &= \frac{1}{2}\langle L|\omega\rangle \\ &= \frac{1}{2}I\langle\omega|\omega\rangle.\end{aligned}$$

The moment of inertia generalizes to the inertia tensor when the assumption of rotation about a fixed *axis* is lifted, allowing the possibility of an instantaneous rotation with angular velocity $\boldsymbol{\omega}$ about a *point*. Let that point be the rigid body's center of mass, where we locate the origin of a system of coordinates. If $\mathbf{r}$ describes the location of dm relative to this origin, this particle's instantaneous velocity relative to it is $\mathbf{v} = \boldsymbol{\omega} \times \mathbf{r}$. Calculate $\mathbf{L}$ again by summing over all the angular momenta of the various particles. Along the way the vector identity

$$\mathbf{A} \times (\mathbf{B} \times \mathbf{C}) = (\mathbf{A} \cdot \mathbf{C})\mathbf{B} - (\mathbf{A} \cdot \mathbf{B})\mathbf{C} \tag{2.7}$$

will prove useful, to write

$$\mathbf{L} = \int \mathbf{r} \times (\boldsymbol{\omega} \times \mathbf{r}) dm \tag{2.8}$$

as

$$\mathbf{L} = \int [(\mathbf{r} \cdot \mathbf{r})\boldsymbol{\omega} - \mathbf{r}(\mathbf{r} \cdot \boldsymbol{\omega})]\, dm. \tag{2.9}$$

Relative to an xyz coordinate system, and noting that

$$\omega^i = \sum_j \omega^j \delta^{ij}, \tag{2.10}$$

the ith component of $\mathbf{L}$ may be written

$$L^i = \sum_j \omega^j \int [\delta^{ij}(\mathbf{r} \cdot \mathbf{r}) - x^i x^j] dm. \tag{2.11}$$

The components of the inertia tensor are the nine coefficients

$$I^{ij} = \int [\delta^{ij}(\mathbf{r} \cdot \mathbf{r}) - x^i x^j] dm. \tag{2.12}$$

Notice that $I^{ij} = I^{ji}$, the inertia tensor is symmetric, and thus it has only six independent components.

If some off-diagonal elements of the inertia tensor are nonzero, then in that coordinate system the angular momentum and the angular velocity are not parallel. Explicitly,

$$L^x = I^{xx}\omega^x + I^{xy}\omega^y + I^{xz}\omega^z \tag{2.13}$$

$$L^y = I^{yx}\omega^x + I^{yy}\omega^y + I^{yz}\omega^z \tag{2.14}$$

$$L^z = I^{zx}\omega^x + I^{zy}\omega^y + I^{zz}\omega^z. \tag{2.15}$$

Eqs. (2.13) - (2.15) can all be summarized as

$$L^i = \sum_{j=1}^{3} I^{ij}\omega^j. \tag{2.16}$$

Throughout this subject we are going to see summations galore like this sum over j. Note that repeated indices are the ones being summed. To make the equations less cluttered, let us agree to drop the capital sigma. With these conventions Eq. (2.16) compactly becomes

$$L^i = I^{ij}\omega^j, \tag{2.17}$$

with the sum over j understood. If for some reason repeated indices are not to be summed in a discussion, the exception will be stated explicitly. For instance, if I wanted to talk about any one of the I^{11}, I^{22}, and I^{33} but without specifying which one, then I would have to say "Consider the I^{kk} (no sum)."

Later on, when summing over repeated indices, we will also impose the stipulation that one of the repeated indices will appear as a superscript and its partner will appear as a subscript. There are reasons for deciding which indices go upstairs and which ones go downstairs, but we need not worry about it for now, because in Euclidean spaces—where we are working at present—the components that carry superscripts and the components labeled with subscripts are redundant. We return to this important issue in chapter 3, when we entertain non-Euclidean spaces and have to distinguish "vectors" from "dual vectors."

2.3 The Electric Quadrupole Tensor

In the 1740s, Benjamin Franklin conducted experiments to show that the two kinds of so-called electrical fire, the "vitreous" and "reninous," were additive inverses of each other. He described experiments in static electricity with glass bottles and wrote, "Thus, the whole force of the bottle, and power of giving a shock, is in the GLASS ITSELF" (Franklin, 1751, original emphasis; see Weart).

Imagine one of Franklin's glass bottles carrying a nonzero electric charge q. Sufficiently far from the bottle the details of its size and shape are indiscernible, and the electric field it produces is indistinguishable from that of a point charge, whose potential varies as the inverse distance, $1/r$. Move closer, and eventually some details of the charge distribution's size and shape begin to emerge. Our instruments will begin to detect nuances in the field, suggesting that the bottle carries some positive and some negative charges that are separated from one another, introducing into the observed field the contribution of an electric dipole (potential $\sim 1/r^2$). Move in even closer, and the distribution of the dipoles becomes apparent, as though the distribution of charge includes pairs of dipoles, a quadrupole (whose potential $\sim 1/r^3$), and so on. The contribution of each multipole can be made explicit by expanding the exact expression for the electric potential in a Taylor series.

Let an infinitesimal charge dq reside at a "source point" located relative to the origin by the vector $\bar{\mathbf{r}} = (\bar{x}, \bar{y}, \bar{z})$. According to Coulomb's law and the superposition principle, the exact electric potential $\phi = \phi(\mathbf{r})$ at position $\mathbf{r}$ is given by

$$\phi(\mathbf{r}) = k \int \frac{dq}{|\mathbf{r} - \bar{\mathbf{r}}|}, \tag{2.18}$$

where $k \equiv \frac{1}{4\pi\epsilon_o}$ denotes the Coulomb constant and

$$dq = \rho(\bar{\mathbf{r}}) d\bar{x} d\bar{y} d\bar{z}. \tag{2.19}$$

Here $\rho(\bar{\mathbf{r}})$ denotes the charge density and $\mathbf{r}-\bar{\mathbf{r}}$ the vector from the source point $\bar{\mathbf{r}}$ to the field point $\mathbf{r}$. Let us assume that the charge density vanishes sufficiently rapidly at infinity for the integral over all space to exist.

Place the origin near the finite charge distribution and perform a Taylor series expansion of $\frac{1}{|\mathbf{r}-\bar{\mathbf{r}}|}$ about $\bar{\mathbf{r}}=\mathbf{0}$. Any function of (x,y,z) expanded about the origin will result in a Taylor series of the form

$$f(x,y,z)=f(0,0,0)+x\left(\frac{\partial f}{\partial x}\right)_0+y\left(\frac{\partial f}{\partial y}\right)_0+z\left(\frac{\partial f}{\partial z}\right)_0+\frac{x^2}{2!}\left(\frac{\partial^2 f}{\partial x^2}\right)_0+\frac{xy}{2!}\left(\frac{\partial^2 f}{\partial x\partial y}\right)_0+\cdots. \tag{2.20}$$

Working in rectanglar coordinates, so that

$$\frac{1}{R}\equiv\frac{1}{|\mathbf{r}-\bar{\mathbf{r}}|}=[(x-\bar{x})^2+(y-\bar{y})^2+(z-\bar{z})^2]^{-1/2}, \tag{2.21}$$

we obtain (recalling the summation convention)

$$\frac{1}{|\mathbf{r}-\bar{\mathbf{r}}|}=\frac{1}{r}+\bar{x}^i\left[\frac{\partial R^{-1}}{\partial\bar{x}^i}\right]_0+\frac{1}{2!}\bar{x}^i\bar{x}^j\left[\frac{\partial^2 R^{-1}}{\partial\bar{x}^i\bar{x}^j}\right]_0+\cdots. \tag{2.22}$$

Evaluating the derivatives and then setting $\bar{x}^i=0$ in them, and putting those results back into Eq. (2.18), develops a multipole expansion as a power series in $1/r$:

$$\phi(\mathbf{r})=\frac{kq}{r}+\frac{k\mu^i}{r^2}p^i+\frac{k}{2r^3}(3\mu^i\mu^j-\delta^{ij})Q^{ij}+\cdots, \tag{2.23}$$

where $\mu^i\equiv x^i/r$ denotes the cosine of the angle between the x^i-axis and $\mathbf{r}$. The distribution's total charge,

$$q=\int dq, \tag{2.24}$$

is the "monopole moment" in the multipole expansion. The second term,

$$p^i\equiv\int\bar{x}^i dq, \tag{2.25}$$

denotes the ith component of the electric dipole moment vector. In the third term, the ijth component of the electric quadrupole tensor introduces itself:

$$Q^{ij}\equiv\int\bar{x}^i\bar{x}^j dq. \tag{2.26}$$

To the extent that the distribution includes pairs of dipoles, it carries a nonzero quadrupole moment, described by the components Q^{ij} of the quadrupole tensor.

Higher-order multipole moments (octupole, 16-pole, and so forth) follow from higher-order terms in the Taylor series. Notice that the net charge carried by a 2^n-pole vanishes for $n>0$. The net electric charge resides in the monopole term, but higher-order multipole moments give increasingly finer details about the charge's spatial distribution.

Multipole expansions can also be written for static magnetic fields; for dynamic electric and magnetic fields including radiation patterns from antennas, atoms, and nuclei; and for gravitational fields.

2.4 The Electromagnetic Stress Tensor

To see another tensor emerge in response to a physics application, let us continue with a distribution of electrically charged particles, but now allow them to move.

Newton's third law says that when particle A exerts a force on particle B, then particle B exerts an equal amount of force, oppositely directed, on particle A. Since force equals the rate of change of momentum, in a two-body-only interaction momentum is exchanged between the bodies but conserved in total. For example, in collisions between billiard balls the momentum exchange is local and immediately apparent.

But suppose particle A is an electron in, say, the antenna of a robot rover scurrying about on the surface of Mars. Let particle B be another electron in a radio receiver located in New Mexico. When electron A aboard the Martian rover jiggles because of an applied variable voltage, it radiates waves in the electromagnetic field. Electron B in New Mexico won't know about A's jiggling for at least 4 minutes (possibly up to 12 minutes, depending on the locations of Mars and Earth in their orbits around the Sun), while the electromagnetic wave streaks from Mars to New Mexico at the speed of light. In the meantime, what of Newton's third law? What of momentum conservation? Electric and magnetic fields carry energy; what momentum do they carry?

To answer our question about the momentum of electromagnetic fields, consider the interaction of charged particles with those fields, by combining Newton's second law with Maxwell's equations.

Newton's second law applies cause-and-effect reasoning to the mechanical motion of a particle of momentum $\mathbf{p}$. The vector sum of the forces $\mathbf{F}$ causes the particle's momentum to change according to

$$\mathbf{F} = \frac{d\mathbf{p}}{dt}. \tag{2.27}$$

Let this particle carry electric charge q, and let other charges in its surroundings produce the electric field $\mathbf{E}$ and the magnetic field $\mathbf{B}$ that exist at our sample particle's instantaneous location. When these are the only forces acting on the sample particle, then Newton's second law applied to it becomes

$$q\mathbf{E} + q(\mathbf{v} \times \mathbf{B}) = \frac{d\mathbf{p}}{dt}. \tag{2.28}$$

To apply Newton's second law to an ensemble of particles, we must sum the set of $\mathbf{F} = d\mathbf{p}/dt$ equations, one for each particle in the ensemble. To do this, let us conceptualize the system of particles as having a charge density ρ. Our sample particle's charge q now gets expressed as an increment of charge $dq = \rho dV$, where dV denotes the infinitesimal volume element it occupies. Moving particles form a current density $\mathbf{j} = \rho \mathbf{v}$, where $\mathbf{v}$ denotes their local velocity. Now Newton's second law, summed over all the particles in the system, reads

$$\int_V dV \left[\rho \mathbf{E} + (\mathbf{j} \times \mathbf{B})\right] = \frac{d\mathbf{p}_m}{dt}, \tag{2.29}$$

where V denotes the volume enclosing the entire distribution of charged particles, and the m subscript on $\mathbf{p}$ denotes the momentum carried by all the system's ponderable matter.

The charge and current densities can be expressed in terms of the fields they produce, thanks to the two inhomogeneous Maxwell equations. They are Gauss's law for the electric field,

$$\nabla \cdot \mathbf{E} = \frac{\rho}{\epsilon_o}, \tag{2.30}$$

and the Ampère-Maxwell law,

$$\nabla \times \mathbf{B} = \mu_o \mathbf{j} + \mu_o \epsilon_o \frac{\partial \mathbf{E}}{\partial t}, \tag{2.31}$$

where $\mu_o \epsilon_o = 1/c^2$. Now Eq. (2.29) becomes

$$\int_V dV \left[\epsilon_o \mathbf{E}\left(\nabla \cdot \mathbf{E}\right) + \frac{1}{\mu_o}(\nabla \times \mathbf{B}) \times \mathbf{B} - \epsilon_o \left(\frac{\partial \mathbf{E}}{\partial t} \times \mathbf{B}\right)\right] = \frac{d\mathbf{p}_m}{dt}. \tag{2.32}$$

The third term on the left-hand side suggests using the product rule for derivatives,

$$\frac{\partial}{\partial t}(\mathbf{E} \times \mathbf{B}) = \left(\frac{\partial \mathbf{E}}{\partial t} \times \mathbf{B}\right) + \left(\mathbf{E} \times \frac{\partial \mathbf{B}}{\partial t}\right), \tag{2.33}$$

and then invoking a third Maxwell equation, the Faraday-Lenz law,

$$\nabla \times \mathbf{E} = -\frac{\partial \mathbf{B}}{\partial t}, \tag{2.34}$$

to replace $\frac{\partial \mathbf{B}}{\partial t}$ with $-\nabla \times \mathbf{E}$. These steps turn Eq. (2.32) into

$$\int_V \left[\epsilon_o \mathbf{E}(\nabla \cdot \mathbf{E}) + \frac{1}{\mu_o}(\nabla \times \mathbf{B}) \times \mathbf{B} - \epsilon_o \frac{\partial}{\partial t}(\mathbf{E} \times \mathbf{B}) - \epsilon_o \mathbf{E} \times (\nabla \times \mathbf{E})\right] = \frac{d\mathbf{p}_m}{dt}. \tag{2.35}$$

To make $\mathbf{E}$ and $\mathbf{B}$ appear as symmetrically as possible, let us add zero in the guise of a term proportional to the divergence of $\mathbf{B}$. This we can do thanks to the remaining Maxwell equation, Gauss's law for the magnetic field:

$$\nabla \cdot \mathbf{B} = 0. \tag{2.36}$$

Now we have

$$\int_V \left[\epsilon_o \mathbf{E}(\boldsymbol{\nabla}\cdot\mathbf{E}) + \frac{1}{\mu_o}\mathbf{B}(\boldsymbol{\nabla}\cdot\mathbf{B}) - \epsilon_o\mathbf{E}\times(\boldsymbol{\nabla}\times\mathbf{E}) - \frac{1}{\mu_o}\mathbf{B}\times(\boldsymbol{\nabla}\times\mathbf{B}) - \epsilon_o\mu_c\frac{\partial}{\partial t}\left(\frac{\mathbf{E}\times\mathbf{B}}{\mu_o}\right)\right] dV = \frac{d\mathbf{p}_m}{dt} \tag{2.37}$$

which looks even more complicated, but will neatly partition itself in a revealing way.

In the last term on the left-hand side, we recognize "Poynting's vector" $\mathbf{S}$, defined according to

$$\mathbf{S} \equiv \frac{\mathbf{E}\times\mathbf{B}}{\mu_o}. \tag{2.38}$$

Since $\mu_o\epsilon_o = 1/c^2$, the last term contains the time derivative of $\mathbf{S}/c^2$. This vector carries the dimensions of momentum density; thus, its volume integral equals the momentum $\mathbf{p}_f$ carried by the fields,

$$\mathbf{p}_f = \int_V \frac{\mathbf{S}}{c^2} dV. \tag{2.39}$$

This rate of change of electromagnetic momentum within V can be transposed from the force side of the equation to join the rate of change of the momentum carried by matter, so that Eq. 2.37 becomes

$$\int_V \left[\epsilon_o\mathbf{E}(\boldsymbol{\nabla}\cdot\mathbf{E}) - \epsilon_o\mathbf{E}\times(\boldsymbol{\nabla}\times\mathbf{E}) + \frac{1}{\mu_o}\mathbf{B}(\boldsymbol{\nabla}\cdot\mathbf{B}) - \frac{1}{\mu_o}\mathbf{B}\times(\boldsymbol{\nabla}\times\mathbf{B})\right] dV = \frac{d}{dt}(\mathbf{p}_m+\mathbf{p}_f) \tag{2.40}$$

To make the components of the electromagnetic stress tensor emerge, write out

$$\mathbf{E}(\boldsymbol{\nabla}\cdot\mathbf{E}) - \mathbf{E}\times(\boldsymbol{\nabla}\times\mathbf{E}) \tag{2.41}$$

by components, and do likewise for $\mathbf{B}$ (see Ex. 2.13). The result gives a component version of Newton's second law for the coupled particle-field system, in which the "electromagnetic stress tensor" T^{ij}, also known as the "Maxwell tensor," makes its appearance explicit,

$$\int_V \frac{\partial T^{ij}}{\partial x^j} dV = \frac{d}{dt}(\mathbf{p}_m + \mathbf{p}_f)^i, \tag{2.42}$$

where

$$T^{ij} \equiv \epsilon_o E^i E^j + \frac{1}{\mu_o} B^i B^j - \eta\delta^{ij}. \tag{2.43}$$

Its components carry the dimensions of pressure or energy density and include the electromagnetic energy density

$$\eta \equiv \frac{1}{2}\epsilon_o E^2 + \frac{1}{2\mu_o}B^2. \tag{2.44}$$

By virtue of Gauss's divergence theorem, the volume integral of the divergence of T^{ij} may be written as the flux of the stress tensor through the closed surface σ enclosing the distribution. Doing this in Eq. (2.42) turns it into

$$\oint_\sigma T^{ij} n^j dA = \frac{d}{dt}(\mathbf{p}_m + \mathbf{p}_f)^i, \tag{2.45}$$

where n^j denotes the jth component of the unit normal vector $\hat{\mathbf{n}}$ that points outward from the surface σ, at the location of the infinitesimal patch of area dA on the surface.

Notice that if σ encloses *all* of space, and because realistic fields vanish at infinity, then the total momentum of the particles and fields is conserved. But if σ refers to a surface enclosing a finite volume, then according to Eq. (2.45), on the left-hand side the flux of energy density or pressure through the closed surface σ equals the rate of decrease of the momentum of particles and fields within the volume enclosed by σ. Conversely, if all the particle and field momentum within V is conserved, whatever volume V happens to be, then Eq. (2.42) becomes an "equation of continuity" expressing the local conservation of energy and momentum:

$$\frac{\partial T^{ij}}{\partial x^j} = 0. \tag{2.46}$$

We could go on with other examples of tensors, but this sample collection of two-index tensors suggests operational descriptions of what they are, in terms of the kinds of tasks they perform. We have seen that, in a given coordinate system, if a two-index tensor has nonzero off-diagonal elements, then a vector stimulus in, say, the x-direction will produce a vector response in the y- and z-directions, in addition to a response in the x direction. This practical reason demonstrates why tensors are useful in physics.

When such directional complexity occurs, one may be motivated to seek a new coordinate system, in terms of which all the off-diagonal components vanish. A procedure for doing this systematically is described in Section 2.6, the "eigenvalue problem." But first let us examine coordinate transformations as they apply to two-index tensors.

2.5 Transformations of Two-Index Tensors

Overheard while waiting outside the open door of a physics professor's office:
Distressed student: "What are these tensors?"
Professor: "Just think of them as matrices."
Greatly relieved student: "Ah,...OK, that I can do."

We have seen that vectors can be represented as column or row matrices, and two-index tensors as square matrices. If our mission consisted of merely doing calcuations with vectors and two-index tensors within a given reference frame, such as computing $K = \frac{1}{2}\omega^i I^{ij} \omega^j$, we could treat our tasks as matrix operations and worry no more about tensor formalities. Are vectors and two-index tensors *merely* matrices? No! There is more to their story.

Whenever we, like the distressed student, merely need numerical results within a given coordinate system, then the calculations do, indeed, reduce to matrix manipulations. The professor in the above conversation was not wrong; she merely told the student what he needed to know to get on with his work at that time. She left for another day an explanation of the relativistic obligations that are laid on tensors, which form an essential part of what they are. Tensors carry an obligation that mere arrays of numbers do not have: they have to transform in specific ways under a change of reference frame.

Before we consider the transformations of tensors *between* reference frames, let us first attempt to write a tensor *itself*, in coordinate-free language, *within* a given reference frame. For an example, consider the inertia tensor. In terms of the coordinates of a given frame its components are

$$I^{ij} = \int dm \left[r^2\delta^{ij} - x^i x^j \right]. \tag{2.47}$$

The idea I am trying to communicate here is that this is a *projection* of the inertia tensor onto some particular xyz coordinate system, not the "tensor itself." A tool for writing the tensor itself in a coordinate-transcending way falls readily to hand with Dirac's bracket notation.

The inertia tensor components are given by $I^{ij} \equiv \langle i|\tilde{\mathbf{I}}|j\rangle$, where the boldface $\tilde{\mathbf{I}}$ distinguishes the "tensor itself" from its components I^{ij}. In addition, the factors in the integrand can be written $r^2 = \mathbf{r}\cdot\mathbf{r} = \langle r|r\rangle$, $x^i = \langle i|r\rangle = \langle r|i\rangle$, and $\langle i|j\rangle = \langle j|i\rangle = \delta^{ij}$. Now I^{ij} may be written

$$\langle i|\tilde{\mathbf{I}}|j\rangle = \int dm\{\langle r|r\rangle\langle i|j\rangle - \langle i|r\rangle\langle r|j\rangle\}. \tag{2.48}$$

Pull the $\langle i|$ and $|j\rangle$ outside the integral to obtain

$$\langle i|\tilde{\mathbf{I}}|j\rangle = \langle i| \int dm\{\langle r|r\rangle\mathbf{1} - |r\rangle\langle r|\}|j\rangle, \tag{2.49}$$

where $\mathbf{1}$ denotes the unit matrix.

We can now identify the inertia tensor itself, $\tilde{\mathbf{I}}$, alternatively denoted by some authors as $\vec{\vec{I}}$ or $\{\mathbf{I}\}$, as the abstract quantity

$$\tilde{\mathbf{I}} \equiv \int dm \left[\, \langle r|r\rangle\mathbf{1} \;-\; |r\rangle\langle r| \,\right]. \tag{2.50}$$

The beauty of this representation lies in the fact that $\tilde{\mathbf{I}}$ can be sandwiched between a pair of unit basis vectors of *any* coordinate system, with a row basis vector $\langle\alpha|$ on the left and a column basis vector $|\beta\rangle$ on the right, to express the $(\alpha\beta)$th component of $\tilde{\mathbf{I}}$ in that basis:

$$\langle\alpha|\tilde{\mathbf{I}}|\beta\rangle \equiv I^{\alpha\beta}. \tag{2.51}$$

Some authors effectively describe this process with a vivid analogy, comparing a two-index tensor to a machine into which one inserts a pair of vectors, like slices of bread dropped into a toaster. But instead of toast, out pops one real number, the tensor component, $\tilde{\mathbf{I}}(\hat{\mathbf{e}}_\alpha, \hat{\mathbf{e}}_\beta) = I^{\alpha\beta}$, another notation for our $\langle\alpha|\tilde{\mathbf{I}}|\beta\rangle = I^{\alpha\beta}$ (more about this way of looking at tensors appears in Ch. 8).

We can change the representation of $\tilde{\mathbf{I}}$ from an old basis (denoted with Latin letters) to a new basis (denoted with Greek letters), by using the completeness relation, which in the old set of coordinates reads

$$|i\rangle\langle i| = \mathbf{1} \tag{2.52}$$

(summed over i). To use this, insert unit matrices into $\langle\alpha|\tilde{\mathbf{I}}|\beta\rangle$ as follows:

$$\begin{aligned}
\langle\alpha|\tilde{\mathbf{I}}|\beta\rangle &= \langle\alpha|\mathbf{1}\ \tilde{\mathbf{I}}\ \mathbf{1}|\beta\rangle \\
&= \langle\alpha|i\rangle\langle i|\tilde{\mathbf{I}}|j\rangle\langle j|\beta\rangle \\
&= \Lambda^{\alpha i} I^{ij} \Lambda^{j\beta} \\
&= \Lambda^{\alpha i} I^{ij} (\Lambda^\dagger)^{\beta j},
\end{aligned}$$

where (recalling primes to emphasize new coordinates vs. unprimed for old ones)

$$\Lambda^{\alpha i} \equiv \langle\alpha|i\rangle = \frac{\partial x'^\alpha}{\partial x^i} \tag{2.53}$$

is the matrix element in the αth row and ith column of the transformation matrix Λ. (These coefficients are not components of tensors, but describe how the coordinates from one system map to those of another system.) These results display a "similarity transformation," which says in our example of the inertia tensor

$$\tilde{\mathbf{I}}' = \Lambda\tilde{\mathbf{I}}\Lambda^\dagger, \tag{2.54}$$

which contains the transformation rule for two-index tensor components: Starting from the matrix multiplication rules, we would write

$$I'^{\alpha\beta} = \Lambda^{\alpha i} I^{ij} \Lambda^{j\beta}, \tag{2.55}$$

which by Eq. (2.53) becomes

$$I'^{\alpha\beta} = \frac{\partial x'^\alpha}{\partial x^i}\frac{\partial x'^\beta}{\partial x^j} I^{\alpha\beta}. \tag{2.56}$$

Let us confirm these ideas explicitly with the inertia tensor. When projected onto some xyz coordinate system, its components are

$$I^{ij} = \int dm \left[\delta^{ij} r^2 - x^i x^j\right]. \tag{2.57}$$

Alternatively, the inertia tensor components could have been evaluated with respect to some other $x'y'z'$ system of coordinates, where the components are

$$I^{i'j'} = \int dm \left[\delta^{i'j'} r'^2 - x'^i x'^j\right], \tag{2.58}$$

which we abbreviate as I'^{ij}. Suppose the xyz coordinates were transformed into those of the $x'y'z'$ system by the orthogonal transformation we saw earlier, which I repeat here for your convenience:

$$x' = x\cos\theta + y\sin\theta \tag{2.59}$$

$$y' = -x\sin\theta + y\cos\theta \tag{2.60}$$

$$z' = z. \tag{2.61}$$

Let us substitute these transformations directly into I'^{ij} and see by a "brute force" calculation how they are related to the I^{ij}. Consider, for example, I'^{xy}:

$$\begin{aligned} I'^{xy} &= -\int dm(x'y') \\ &= -\int dm(x\cos\theta + y\sin\theta)(-x\sin\theta + y\cos\theta) \\ &= \int dm[(x^2-y^2)\cos\theta\sin\theta + xy(\sin^2\theta - \cos^2\theta)] \\ &= \cos\theta\sin\theta(I^{yy} - I^{xx}) - I^{xy}(\sin^2\theta - \cos^2\theta). \end{aligned}$$

Let us see whether this result agrees with the transformation rule of Eq. (2.56). Writing out the right-hand side of what Eq. (2.56) instructs us to do, we form the sum

$$I'^{xy} = \frac{\partial x'}{\partial x}\frac{\partial y'}{\partial x}I^{xx} + \frac{\partial x'}{\partial x}\frac{\partial y'}{\partial y}I^{xy} + \frac{\partial x'}{\partial y}\frac{\partial y'}{\partial x}I^{yx} + \frac{\partial x'}{\partial y}\frac{\partial y'}{\partial y}I^{yy}. \tag{2.62}$$

From the foregoing transformation equations we evaluate the partial derivatives and obtain

$$I'^{xy} = [-\cos\theta\sin\theta]I^{xx} + [\cos^2\theta]I^{xy} + [-\sin^2\theta]I^{yx} + [\sin\theta\cos\theta]I^{yy}, \tag{2.63}$$

which is identical to our "brute force" calculation.

A pattern seems to be emerging in how tensors whose components carry various numbers of indices transform under coordinate transformations. For a scalar λ,

$$\lambda' = \lambda. \tag{2.64}$$

Scalars are invariant, carry no coordinate indices, and in the context of tensor analysis are called "tensors of order 0." For a vector with components A^i,

$$A'^i = \frac{\partial x'^i}{\partial x^j}A^j, \tag{2.65}$$

which transform like coordinate displacements themselves, carry one index, and are "tensors of order 1." For a two-index, or "order- 2," tensor,

$$T'^{ij} = \frac{\partial x'^i}{\partial x^k}\frac{\partial x'^j}{\partial x^n}T^{kn}. \tag{2.66}$$

The number of partial derivative factors in the transformation rule equals the number of indices, the tensor's order. Extensions to tensors of higher order will be forthcoming.

In response to the question of what a tensor *is* (aside from how it *behaves* under changes of coordinates), we now have at least one answer in terms of a coordinate-free mathematical entity. For the inertia tensor, it is the operator $\tilde{\mathbf{I}}$ of Eq. (2.50). For the purposes of calculating its components, merely sandwich the operator between a pair of unit basis vectors to generate the nine components (in three-dimensional space) of the inertia tensor, represented by a square matrix. But in so doing we must remember that inertia tensors—indeed, all tensors—are not *just* matrices. Tensors also have the obligation to respect precise rules for how their components transform under a change of coordinate system.

Changing a coordinate system can be an effective strategy for solving a problem. A problem that looks complicated in one reference frame may appear simple in another one. If a two-index tensor has nonzero off-diagonal components, it may be because the object or interaction it describes is genuinely complicated. But it may also be due to our evaluating its components with respect to a set of axes that make the tensor *appear* unnecessarily complicated. A systematic method for finding a new coordinate system, in terms of which all the off-diagonal components vanish in the matrix representation, forms the "eigenvector and eigenvalue" problem, to which we turn next.

2.6 Finding Eigenvectors and Eigenvalues

If you have ever driven a car with unbalanced wheels, you know that the resulting shimmies and shakes put unnecessary wear and tear on the car's suspension system and tires. This situation can become dangerous when the shaking of an unbalanced wheel goes into a resonance mode.

Balanced or not, the tire/wheel assembly is supposed to spin about a fixed axis. Let that axis be called the z-axis. The xy axes that go with it lie in a plane perpendicular to the axle. Although the wheel's angular velocity points along the z axis, an unbalanced wheel's angular momentum points slightly off the z axis, sweeping out a wobbly cone around it. Because $|L\rangle = \tilde{\mathbf{I}}|\omega\rangle$, the shaking about of an unbalanced spinning wheel/tire assembly suggests that its inertia tensor has nonzero off-diagonal terms relative to these xyz axes. When a mechanic balances the wheel, small weights are attached to the wheel rim. The mass distribution of the wheel/tire assembly is modified from its previous state so that, when evaluated with respect to the original coordinate system, the new inertia tensor has become diagonal. Then the spinning wheel's angular momentum vector lines up with the angular velocity vector. The ride is much smoother, the tire does not wear out prematurely, and vibration resonances do not occur.

The mechanic had to change the mass distribution because changing the spin axis was not an option. However, from a mathematical point of view we can keep the original mass distribution, but find a new coordinate system in terms of which all the off-diagonal terms in the matrix representation of the tensor are zero. Then the nonzero components all reside along the diagonal and are called "eigenvalues." The basis vectors of the new coordinate system, with respect to which the tensor is now diagonal, are called the "eigenvectors" of the tensor. Each eigenvector has its corresponding eigenvalue.

To find the eigenvectors and eigenvalues, one begins with the tensor expressed in terms of the original coordinates x^i, where the tensor has components $I^{ij} = \langle i|\tilde{\mathbf{I}}|j\rangle$. The task that lies ahead is to construct a new set of coordinates x^α with their basis vectors $|\alpha\rangle$, the eigenvectors, such that $I^{\alpha\beta} = \langle\alpha|\tilde{\mathbf{I}}|\beta\rangle$ is diagonal,

$$I^{\alpha\beta} = \lambda\delta^{\alpha\beta}, \tag{2.67}$$

where λ, the eigenvalue, corresponds to an eigenvector $|\beta\rangle$, and $\delta^{\alpha\beta}$ is the Kronecker delta. Eq. (2.67) is a component version of the matrix equation

$$\tilde{\mathbf{I}}|\beta\rangle = \lambda|\beta\rangle, \tag{2.68}$$

which can be transposed as

$$(\tilde{\mathbf{I}} - \lambda\mathbf{1})|\beta\rangle = |0\rangle, \tag{2.69}$$

where $\mathbf{1}$ denotes the unit matrix and $|0\rangle$ the zero vector.

One obvious solution of this equation is the unique but trivial one, where all the $|\beta\rangle$ are zero vectors. Not very interesting! To find the nontrivial solution, we must invoke the "theorem of alternatives" (also called the "invertible matrix theorem") that comes from linear algebra. Appendix B presents a statement of the theorem of alternatives. I suppose the theorem takes the name "alternatives" from the observation that the determinant $|M|$ of a matrix either equals zero or does not equal zero. The theorem says that for a known matrix M and an unknown vector $|x\rangle$, if $|M| \neq 0$, then the solution of the homogeneous matrix equation $M|x\rangle = |0\rangle$ is the unique but trivial $|x\rangle = |0\rangle$, but if $|M| = 0$, then a nontrivial–but non-unique–solution exists. The nontrivial solution can be made unique after the fact, by imposing the supplementary requirement that the eigenvectors are constructed to be mutually orthogonal *unit* vectors, so that we require

$$\langle\alpha|\beta\rangle = \delta^{\alpha\beta}. \tag{2.70}$$

Then the eigenvectors will form an orthonormal basis for a new coordinate system.

Guided by the theorem of alternatives, we set to zero the determinant

$$|\tilde{\mathbf{I}} - \lambda\mathbf{1}| = 0, \tag{2.71}$$

which yields a polynomial equation for λ whose roots are the eigenvalues. Once we have the eigenvalues, we can find the eigenvectors from Eq. (2.68) and the supplementary orthonormality conditions. An example may illustrate the procedure better than speaking in generalities.

Consider an inertia tensor evaluated in some xyz coordinate system where its array of components has the form

$$I^{ij} = \begin{pmatrix} A & -C & 0 \\ -C & B & 0 \\ 0 & 0 & A+B \end{pmatrix}. \tag{2.72}$$

Such an inertia tensor applies to a piece of thin sheet metal lying in the xy plane. If the sheet metal is a square of mass m with sides of length a, and the edges lie along the xy axes with a corner at the origin, then $A = B = \frac{1}{3}ma^2$ and $C = -\frac{1}{4}ma^2$. For this case the determinant becomes

$$\begin{vmatrix} \frac{1}{3}\eta - \lambda & -\frac{1}{4}\eta & 0 \\ -\frac{1}{4}\eta & \frac{1}{3}\eta - \lambda & 0 \\ 0 & 0 & \frac{2}{3}\eta - \lambda \end{vmatrix} = 0, \tag{2.73}$$

where $\eta \equiv ma^2$. The roots are

$$\lambda = \frac{8}{12}\eta,\ \frac{7}{12}\eta,\ \frac{1}{12}\eta. \tag{2.74}$$

Now we can find the eigenvectors one at a time. Let us start with $\lambda = \frac{8}{12}\eta$. Put it back into Eq. (2.68), and parameterize the eigenvector as

$$|\beta\rangle = \begin{pmatrix} \beta^1 \\ \beta^2 \\ \beta^3 \end{pmatrix} \tag{2.75}$$

so that Eq. (2.68) says

$$\frac{1}{12}\eta \begin{pmatrix} 4 & -3 & 0 \\ -3 & 4 & 0 \\ 0 & 0 & 8 \end{pmatrix} \begin{pmatrix} \beta^1 \\ \beta^2 \\ \beta^3 \end{pmatrix} = \frac{8}{12}\eta \begin{pmatrix} \beta^1 \\ \beta^2 \\ \beta^3 \end{pmatrix}. \tag{2.76}$$

Component by component this matrix equation is equivalent to the set of relations

$$\begin{aligned} 4\beta^1 &= -3\beta^2 \\ 3\beta^1 &= -4\beta^2 \\ \beta^3 &= \beta^3, \end{aligned}$$

which yields $\beta^1 = \beta^2 = 0$ with β^3 left undetermined. All we know so far about $|\beta\rangle$ is that is has only a z-component when mapped in the original xyz axes. This is where the "non-uniquness" promised by the theorem of alternatives shows itself. To nail $|\beta\rangle$ down, let us choose to make it a unit vector, so that $\langle\beta|\beta\rangle = 1$, which gives $\beta^3 = 1$. Now we have a unique eigenvector corresponding to the eigenvalue $8\eta/12$.

In a similar manner the other two eigenvectors are found. Give new names $|\rho\rangle$, $|\sigma\rangle$, and $|\zeta\rangle$ to the eigenvectors. In terms of the original xyz axes, the eigenvector corresponding to the eigenvalue $\lambda = \frac{1}{12}\eta$ takes the form

$$|\rho\rangle = \frac{1}{\sqrt{2}}\begin{pmatrix} 1 \\ 1 \\ 0 \end{pmatrix}. \tag{2.77}$$

For the eigenvalue $\lambda = \frac{7}{12}\eta$,

$$|\sigma\rangle = \frac{1}{\sqrt{2}}\begin{pmatrix} 1 \\ -1 \\ 0 \end{pmatrix}, \tag{2.78}$$

and for the eigenvector that has $\lambda = \frac{8}{12}\eta$,

$$|\zeta\rangle = \begin{pmatrix} 0 \\ 0 \\ 1 \end{pmatrix}. \tag{2.79}$$

If these eigenvectors are drawn in the original xyz coordinate system, $|\zeta\rangle$ coincides with the original unit vector $\hat{\mathbf{k}}$, $|\rho\rangle = \frac{1}{\sqrt{2}}(\hat{\mathbf{i}}+\hat{\mathbf{j}})$, and $|\sigma\rangle = \frac{1}{\sqrt{2}}(\hat{\mathbf{i}}-\hat{\mathbf{j}})$. These eigenvectors form a set of basis vectors for a new set of $x'y'z'$ axes that are rotated about the original z axis through -45 degrees. Notice that the eigenvector $|\rho\rangle$ points along the square sheet metal's diagonal, one of its symmetry axes.

Now recompute the components of the inertia tensor in this new $\{|\rho\rangle, |\sigma\rangle, |\zeta\rangle\}$ basis, and denote them as $I^{\rho\sigma} = \langle\rho|\tilde{\mathbf{I}}|\sigma\rangle$ and so forth. We find that there are no nonzero off-diagonal elements, and the eigenvalues occupy positions along the diagonal:

$$\begin{aligned} I^{\mu\nu} &= \begin{pmatrix} I^{\rho\rho} & I^{\rho\sigma} & I^{\rho\zeta} \\ I^{\sigma\rho} & I^{\sigma\sigma} & I^{\sigma\zeta} \\ I^{\zeta\rho} & I^{\zeta\sigma} & I^{\zeta\zeta} \end{pmatrix} \\ &= \frac{\eta}{12}\begin{pmatrix} 1 & 0 & 0 \\ 0 & 7 & 0 \\ 0 & 0 & 8 \end{pmatrix}. \end{aligned}$$

If the piece of sheet metal were to be set spinning about any one of the three axes defined by the eigenvectors, then its angular momentum would be parallel to the angular velocity, and the moments of inertia would be the eigenvalues.

One complication occurs whenever two or more eigenvalues happen to be identical. In that case the system is said to be "degenerate." The eigenvectors of degenerate eigenvalues can then be *chosen*, so long as each one is orthogonal to all the other eigenvectors. In the example of the balanced wheel, the eigenvector along the spin axis has a distinct eigenvalue, but the eigenvalues of the other two eigenvectors are degenerate, because those eigenvectors could point anywhere in a plane parallel to the wheel's diameter. One chooses them to be orthonormal so they can serve as a normalized basis. A balanced wheel is highly symmetrical. Degenerate eigenvalues correspond to a symmetry of the system.

If the eigenvalue and eigenvector results are compared to calculations of moments of inertia for simple rigid bodies rotating about a fixed axis—the calculations familiar from introductory physics—it will be noticed that, in most instances, the object was symmetrical about one or more axes: cylinders, spheres, rods, and so forth. Those moments of inertia, evaluated about a fixed axis of symmetry, were the eigenvalues of an inertia tensor, and the axis about which I was computed happened to be colinear with one of the eigenvectors.

2.7 Two-Index Tensor Components as Products of Vector Components

A two-index tensor's components transform as

$$T'^{ij} = \frac{\partial x'^i}{\partial x^k}\frac{\partial x'^j}{\partial x^n}T^{kn}. \tag{2.80}$$

On the other hand, the product of two vector components transforms as

$$\begin{aligned} A'^i B'^j &= \left(\frac{\partial x'^i}{\partial x^k}A^k\right)\left(\frac{\partial x'^j}{\partial x^n}B^n\right) \\ &= \frac{\partial x'^i}{\partial x^k}\frac{\partial x'^j}{\partial x^n}A^k B^n. \end{aligned}$$

This appears to be the same transformation rule as the component of a second-order tensor, suggesting that we could rename $A^k B^n \equiv T^{kn}$.

Here we find another way to appreciate what a tensor "is" besides defining it in terms of how it transforms: a tensor component with n indices is equivalent to a product of n vector components. We have already seen this to be so for the inertia tensor components I^{ij}, and Ex. 2.17 shows it to be so for the susceptibility tensor components χ^{ij}.

Incidentally, a significant result can now be demonstrated through the "back door" of the inertia tensor. Because the inertia tensor is a sum, we can break it into two chunks,

$$I^{ij} = \delta^{ij}\int r^2 dm - \int r^i r^j dm. \tag{2.81}$$

Since the I^{ij} and the $\int r^i r^j dm$ are components of second-order tensors, and $\int r^2 dm$ is a scalar, it follows that the Kronecker delta, δ^{ij}, is a second-order tensor.

2.8 More Than Two Indices

In the derivation of the quadrupole tensor of Section 2.3, it was suggested that higher-order multipoles could be obtained if more terms were developed in the Taylor series expansion of $1/|\mathbf{r}-\bar{\mathbf{r}}|$, where $\bar{\mathbf{r}}$ identifies the source point and $\mathbf{r}$ the field point. When the multipole expansion for the electric potential includes the term after the quadrupole, the $1/r^4$ octupole term appears:

$$\phi_{octupole} = \frac{1}{2r^4}\left[5\mu^i\mu^j\mu^k - \delta^{ij}\mu^k - \delta^{jk}\mu^i - \delta^{ki}\mu^j\right]\Omega^{ijk}, \tag{2.82}$$

where $\mu^i = x^i/r$ and

$$\Omega^{ijk} \equiv \int \bar{x}^i\bar{y}^j\bar{z}^k\,dq. \tag{2.83}$$

Is Ω^{ijk} a third-order tensor? The transformations of the source point coordinates, such as

$$\bar{x'}^i = \frac{\partial\bar{x'}^i}{\partial\bar{x}^j}\bar{x}^j, \tag{2.84}$$

do indeed identify Ω^{ijk} as a tensor of order 3, because the product of coordinates in Eq. (2.83) gives

$$\Omega'^{ijk} = \frac{\partial\bar{x'}^i}{\partial\bar{x}^l}\frac{\partial\bar{x'}^j}{\partial\bar{x}^m}\frac{\partial\bar{x'}^k}{\partial\bar{x}^n}\Omega^{lmn}. \tag{2.85}$$

As implied by the multipole expansion, tensors of arbitrarily high order exist in principle. We will encounter an important fourth-order tensor, the Riemann tensor, in Chapter 5.

2.9 Integration Measures and Tensor Densities

Hang on a minute! Haven't we been overlooking something? The inertia tensor components I^{ij} were defined by an integral whose measure was a mass increment dm. The quadrupole and octupole tensors Q^{ij} and Ω^{ijk} included the integration measure dq. But to evaluate the integrals for a specific mass or charge distribution, we must write dm or dq as a density ρ multiplied by a volume element (dropping overbars for source points, since we are not involving field points in this discussion): dq or $dm = \rho dxdydz \equiv \rho d^3x$, which introduces more displacements. Shouldn't we take into account the tranformation behavior of this integration measure ρd^3x, which includes a product of displacements? Yes, we should. However, it is all right so far with our inertia tensor and electric multipole moments because dm and dq are scalars. Evidently some non-scalar piece of ρ "cancels out" any non-scalar behavior of $dxdydz$, leaving ρd^3x a scalar.

This leaves us free to determine the tensor character of the inertia tensor or multipole moments from the remaining factors (which are coordinates) that appear behind the integral sign.

However, this brings up a deeper issue. If we did not have the charge or mass density in the integral, then we *would* have to consider the transformation properties of $dxdydz$ when studying the transformation behavior of something like I^{ij} or Ω^{ijk}. This reveals the distinction between "tensors" and "tensor densities," which will be pursued further in Section 3.7. For example, the Kronecker delta δ^{ij} is a tensor, but the Levi-Civita symbol ϵ^{ijk} is a tensor density and not, in general, a tensor.

The Levi-Civita symbol offers an interesting study. Under some transformations it can be considered to be a tensor, but not under all transformations. The situation depends on whether the transformation is a "proper" or an "improper" one. This distinction has to do with the determinant of the matrix of transformation coefficients. Recall that, in matrix language, the coordinates change according to

$$|x'\rangle = \Lambda|x\rangle, \tag{2.86}$$

where, to maintain $\langle x'|x'\rangle = \langle x|\Lambda^\dagger\Lambda|x\rangle = \langle x|x\rangle$, the transformation matrix Λ satisfies the unitarity condition,

$$\Lambda^\dagger\Lambda = \mathbf{1}. \tag{2.87}$$

Since the determinant of a matrix and the determinant of its adjoint are equal, taking the determinant of the unitarity condition yields

$$|\Lambda|^2 = 1 \tag{2.88}$$

and thus

$$|\Lambda| = \pm 1. \tag{2.89}$$

"Proper transformations," such as orthogonal rotations of axes, have $|\Lambda| = +1$. Transformations with $|\Lambda| = -1$ are called "improper" and include inversions of the coordinate axes, $x' = -x, y' = -y, z' = -z$. If the xyz axes are right-handed ($\hat{\mathbf{i}}\times\hat{\mathbf{j}} = \hat{\mathbf{k}}$), then the inverted $x'y'z'$ system is left-handed ($\hat{\mathbf{i}}'\times\hat{\mathbf{j}}' = -\hat{\mathbf{k}}'$). The Levi-Civita symbol ϵ^{ijk} in Cartesian coordinates transforms as a tensor of order 3 under orthogonal transformations for which $|\Lambda| = +1$. But ϵ^{ijk} does *not* transform as a tensor under reflection, for which $|\Lambda| = -1$. Such an object is is called a "pseudotensor." It is also called a "tensor density of weight $+1$," as will be discussed in Section 3.7.

2.10 Discussion Questions and Exercises

Discussion Questions

Q2.1 A set of 2-index tensor components can be represented as a matrix and stored as an array of numbers written on a card. How could a 3-index tensor's components be represented in hard-copy storage? What about a 4-index tensor?

Q2.2 For tensor components bearing two or more indices, the order of the indices matters in general. However, some symmetries or asymmetries in a tensor's structure mean that the interchange of indices reduces the number of independent components. In three-dimensional space, how many independent components exist when a two-index tensor is symmetric, $T^{ij} = T^{ji}$? How many independent components exist for an antisymmetric two-index tensor, $F^{ij} = -F^{ji}$? What are the answers to these questions in four dimensions?

Q2.3 Comment on similarities and differences between the inertia tensor and the quadrupole tensor.

Q2.4 An electric monopole—a point charge—located at the origin has no dipole moment, and an ideal dipole, made of two oppositely charged point charges, has no monopole moment if one of its two charges sits at the origin. However, even an ideal dipole has a quadrupole moment. An ideal quadrupole has no monopole or dipole moment, but it may have octupole and higher moments in addition to its quadrupole moment. This pattern continues with other multipoles. What's going on here? If a monopole does not sit at the origin, does it acquire a dipole moment?

Q2.5 Discuss whether it makes sense to write the inertia tensor components (or any other two-index tensor) as a *column* vector whose entries are *row* vectors:

$$\{I^{ij}\} = \begin{pmatrix} \langle a| \\ \langle b| \\ \langle c| \end{pmatrix}. \tag{2.90}$$

Would it make sense to think of the array of I^{ij} as forming a row vector whose entries are column vectors, such as

$$\{I^{ij}\} = (|a\rangle|b\rangle|c\rangle)? \tag{2.91}$$

Q2.6 In rectangular coordinates, the z-component of the angular mometum vector $\mathbf{L}$ may be written

$$L^z = xp^y - yp^x \tag{2.92}$$

or as

$$L^3 = x^1p^2 - x^2p^1 \equiv L^{12}, \tag{2.93}$$

where $\mathbf{p}$ denotes the linear momentum. In other words, for ijk denoting the numerals 123 in cyclic order,

$$L^i = x^jp^k - x^kp^j \equiv L^{jk}. \tag{2.94}$$

Can we think of angular momentum as a second-order tensor instead of a vector? Would your conclusion hold for any cross product $\mathbf{A} \times \mathbf{B}$?

Q2.7 Why are the dot and cross products typically defined in rectangular coordinates? Do we lose generality by doing this?

Q2.8 In nuclear physics, the interaction between the proton and neutron in the deuteron (the nucleus of hydrogen-2) includes a so-called tensor interaction, proportional to

$$3(\mathbf{S}_p \cdot \hat{\mathbf{r}})(\mathbf{S}_n \cdot \hat{\mathbf{r}}) - \mathbf{S}_p \cdot \mathbf{S}_2 \tag{2.95}$$

where $\mathbf{S}_n$ and $\mathbf{S}_p$ are the spin of the neutron and the proton respectively, and $\hat{\mathbf{r}}$ is the unit vector in the radial direction of spherical coordinates. Why do you suppose this interaction is called the potential corresponding to a "tensor force"? (See Roy and Nigam, pp. 72, 78.)

Q2.9 The inverse distance $|\mathbf{r} - \bar{\mathbf{r}}|^{-1}$ from the source point $\bar{\mathbf{r}}$ to field point $\mathbf{r}$ can be expanded, in spherical coordinates, as a superposition of "spherical harmonics" $Y_{lm}(\theta, \phi)$, where $l = 0, 1, 2, 3, ...$ and $m = -l, -(l-1), ..., l-1, l$ (see any quantum mechanics or electrodynamics textbook for more on the spherical harmonics). The spherical harmonics form an orthonormal basis on the surface of a sphere; any function of latitude and longitude can be expressed as a superposition of them. A theorem (see Jackson; the subscripts on Y_{lm} are Jackson's notation) shows that

$$\frac{1}{|\mathbf{r} - \bar{\mathbf{r}}|} = 4\pi \sum_{l=0}^{\infty} \sum_{m=-l}^{l} \frac{1}{2l+1} \frac{r_<^l}{r_>^{l+1}} Y_{lm}^*(\bar{\theta}, \bar{\phi}) Y_{lm}(\theta, \phi), \tag{2.96}$$

where $r_<$ is the lesser, and $r_>$ the greater, of $|\mathbf{r}|$ and $|\bar{\mathbf{r}}|$, and $*$ means complex conjugate. With this, the electrostatic potential may be written

$$\phi(\mathbf{r}) = 4\pi k \sum_{l=0}^{\infty} \sum_{m=-l}^{+l} q^{lm} \frac{Y_{lm}(\theta, \phi)}{r^{l+1}}, \tag{2.97}$$

where

$$q^{lm} \equiv \int Y_{lm}^*(\bar{\theta}, \bar{\phi}) \bar{r} \rho(\bar{\mathbf{r}}) d^3\bar{x} \tag{2.98}$$

denotes the "multipole moment of order lm" in spherical coordinates. J. D. Jackson remarks on "the relationship of the Cartesian multipole moments [such as our Q^{ij}] to the spherical multipole moments. The former are $(l-1)(l+2)/2$ in number and for $l > 1$ are more numerous than the $(2l+1)$ spherical components. There is no contradiction here. The root of the differences lies in the different rotational transformation properties of the two types of multipole moments." He then refers the reader to one of his exercises. What is going on here? (See Jackson, pp. 102, 136-140, and his Ex. 4.3 on p. 164.)

Q2.10 Multipole moments and components of the inertia tensor involve volume integrals over a mass density, so that dm or $dq = \rho d^3x$. How do we handle these integrals when the mass or charge distribution happens to be one or more distinct point sources? What about infinitesimally thin lines or sheets? If you have not met it already, find out whatever you can about the "Dirac delta function," essentially the density of a point source—it is a singular object, but its integral gives a finite result.

Exercises

2.1 The Newtonian gravitational field $\mathbf{g}$ produced by a massive object is the negative gradient of a potential ϕ, so that $\mathbf{g} = -\nabla\phi$, where

$$\phi(\mathbf{r}) = -G \int \frac{\rho(\bar{\mathbf{r}})}{|\mathbf{r} - \bar{\mathbf{r}}|} d\bar{x} d\bar{y} d\bar{z}, \tag{2.99}$$

G is Newton's gravitational constant, and ρ is the object's mass density.
(a) Place the origin of the coordinate system at the object's center of mass, and perform a multipole expansion at least through the quadrupole term.
(b) Show that the gravitational dipole moment vanishes with the object's center of mass located at the origin.
(c) Show that the dipole moment vanishes even if the object's center of mass is displaced from the origin. Thus, the first correction beyond a point mass appears in the quadrupole term.
(d) What is different about Coulomb's law and Newton's law of universal gravitation that precludes a Newtonian gravitational field from having a dipole moment?

2.2 Fill in the steps in the derivation of the electric quadrupole tensor in rectangular coordinates.

2.3 (a) Transform the electric quadrupole tensor from rectangular to spherical coordinates.
(b) Write the electrostatic potential in a multipole expansion using the law of cosines in spherical coordinates to express the angle θ (see Appendix A) in terms of r and $\bar{r}$. Show that (see Griffiths, p. 147)

$$\phi(\mathbf{r}) = k \sum_{n=0}^{\infty} \frac{Q^n}{r^{n+1}}, \tag{2.100}$$

where in terms of Legendre polynomials $P_n(x)$ (see any electricity and magnetism or quantum text)

$$Q^n = \int \bar{r}^n P_n(\cos\bar{\theta})\rho(\bar{r})d^3\bar{r} \tag{2.101}$$

(no sum over n in the integral). Can we say that Q^n is a tensor?

2.4 Calculate the inertia tensor for a solid cylinder of radius R, height h, uniform density, and mass m. Use a rectangular or cylindrical system of coordinates, with the z axis passing through the cylinder's symmetry axis. Comment on what you find for the off-diagonal elements. With these same axes, compare the diagonal elements to various moments of inertia for the cylinder as done in Introductory Physics.

2.5 Two uniform line charges of length $2a$, one carrying uniformly distributed charge q and the other carrying charge $-q$, cross each other in such a way that their endpoints lie at $(\pm a, 0, 0)$ and $(0, \pm a, 0)$. Determine the electric potential $\phi(x, y, z)$ for field points $|\mathbf{r}| \gg a$, out through the quadrupole term (adapted from Panofsky and Phillips, p. 27).

2.6 Compute the inertia and quadrupole tensors for two concentric charged rings of radii a and b, with $b > a$, where each ring carries uniform densities of charge and mass. Let both rings have mass m, but the inner ring carrry charge $+q$ and the outer one carry charge $-q$ (adapted from Panofsky and Phillips, p. 27).

2.7 When a charge distribution is symmetric about the z-axis, show that the quadrupole contribution to the electrostatic potential simplifies to

$$\phi_{quadrupole} = k\frac{1}{4r^3}(3\cos^2\theta - 1)(3Q_{33} - TrQ), \tag{2.102}$$

where θ denotes the "latitude" angle measured from the $+z$-axis, $k = \frac{1}{4\pi\epsilon_o}$, and the trace of the quadrupole tensor, $\mathrm{Tr}Q$, is the sum of the diagonal elements, $\mathrm{Tr}Q \equiv Q^{11} + Q^{22} + Q^{33} = Q^{ii}$.

2.8 Consider a uniformly charged spherical shell of radius R which carries uniformly distributed total charge q. Calculate the force exerted by the southern hemisphere on the northern hemisphere (see Griffiths, p. 353).

2.9 Using xyz axes running along a cube's edges, with origin at the cube's corner, show that the inertia tensor of a cube of mass m is

$$I^{ij} = \frac{ma^2}{12}\begin{pmatrix} 8 & -3 & -3 \\ -3 & 8 & -3 \\ -3 & -3 & 8 \end{pmatrix}, \tag{2.103}$$

where a denotes the length of the cube's sides (see Marion and Thornton, p. 418).

2.10 Find the eigenvalues and eigenvectors for the cube whose inertia tensor is originally given by the matrix representation of Ex. 2.9.

2.11 Another language for dealing with second-order tensors is found in "dyads." Dyads are best introduced with an example. Consider two vectors in three-dimensional Euclidean space,

$$\mathbf{A} = 3\hat{\mathbf{i}} + 2\hat{\mathbf{j}} + 5\hat{\mathbf{k}} \tag{2.104}$$

and

$$\mathbf{B} = 7\hat{\mathbf{i}} + 11\hat{\mathbf{k}}. \tag{2.105}$$

Define their *dyad*, denoted **AB**, by distributive multiplication. This is not a dot product multiplication (which gives a scalar), or a cross product multiplication (which gives a vector perpendicular to **A** and **B**), but something else:

$$\begin{aligned}\mathbf{AB} &\equiv (3\hat{\mathbf{i}} + 2\hat{\mathbf{j}} + 5\hat{\mathbf{k}})(7\hat{\mathbf{i}} + 0\hat{\mathbf{j}} + 11\hat{\mathbf{k}}) \\ &= 21\hat{\mathbf{i}}\hat{\mathbf{i}} + 0\hat{\mathbf{i}}\hat{\mathbf{j}} + 33\hat{\mathbf{i}}\hat{\mathbf{k}} \\ &+ 14\hat{\mathbf{j}}\hat{\mathbf{i}} + 0\hat{\mathbf{j}}\hat{\mathbf{j}} + 22\hat{\mathbf{j}}\hat{\mathbf{k}} \\ &+ 35\hat{\mathbf{k}}\hat{\mathbf{i}} + 0\hat{\mathbf{k}}\hat{\mathbf{j}} + 55\hat{\mathbf{k}}\hat{\mathbf{k}}.\end{aligned}$$

A dyad can be displayed as a matrix, which in our example is

$$\mathbf{AB} = \begin{pmatrix} 21 & 0 & 33 \\ 14 & 0 & 22 \\ 35 & 0 & 55 \end{pmatrix}. \tag{2.106}$$

(a) Does $\mathbf{AB} = \mathbf{BA}$?
A dyad starts making sense when it gets multiplied by a third vector with a scalar product. For example, along comes the vector $\mathbf{C} = 2\hat{\mathbf{i}} + 3\hat{\mathbf{j}} + 4\hat{\mathbf{k}}$. **C** hooks onto the dyad, either from the left or from the right, through dot product multiplication, where $\mathbf{C} \cdot \mathbf{AB}$ means $(\mathbf{C} \cdot \mathbf{A})\mathbf{B}$. For instance, the leading term in $\mathbf{AB} \cdot \mathbf{C}$ is $21\hat{\mathbf{i}}\hat{\mathbf{i}} \cdot \mathbf{C} = 42\hat{\mathbf{i}}$.
(b) Calculate $\mathbf{AB} \cdot \mathbf{C}$.
(c) Does $\mathbf{C} \cdot \mathbf{AB} = \mathbf{AB} \cdot \mathbf{C}$?
(d) Write the inertia tensor of Ex. 2.9 in dyad notation. How would the relation between angular momentum and angular velocity be expressed in terms of dyads, when these two vectors are not parallel?
(e) Write the quadrupole tensor in dyad notation and in the Dirac notation (in the latter case, as we did for $\tilde{\mathbf{I}}$).
(f) Can $\mathbf{AB} \times \mathbf{C}$ be defined?

2.12 A superposition of dyads is called a "dyadic." (a) Write the completeness relation as a dyadic, and (b) use it to express the synthesis/analysis relations for a vector.

2.13 (a) In the derivation of the electromagnetic stress tensor, show that

$$E^i(\nabla \cdot \mathbf{E}) + [\mathbf{E} \times (\nabla \times \mathbf{E})]^i = \frac{\partial}{\partial x^j}[E^i E^j - \frac{1}{2}(\mathbf{E} \cdot \mathbf{E})\delta^{ij}]. \tag{2.107}$$

Eq. (1.116) may be useful.
(b) $\frac{\partial T^{ij}}{\partial x^j}$ was said to be the divergence of the tensor T^{ij}. Since we are used to thinking of "the divergence" as operating on a vector, how can this claim be justified?

2.14 Suppose a fluid flows over a surface, such as air flowing over an airplane wing. The force increment that the wing exerts on a patch of fluid surface of area dA may be written in terms of the components of a hydrodynamic stress tensor T^{ij} (e.g., see Acheson, pp. 26-27, 207):

$$dF^i = T^{ij} n^j dA. \tag{2.108}$$

where the normal unit vector $\hat{\mathbf{n}}$ is perpendicular to the surface element of area dA and points outward from closed surfaces. The tangential component ("shear") is due to viscosity; the normal component is the pressure. A simple model of viscosity, a "Newtonian fluid," models the viscous force as proportional to the velocity gradient, so that

$$T^{ij} = -\delta^{ij} P + \eta \left(\frac{\partial v^i}{\partial x^j} + \frac{\partial v^j}{\partial x^i} \right) = T^{ji}, \tag{2.109}$$

where η denotes the coefficient of viscosity and v^i a component of fluid velocity. When gravity is taken into account, Newton's second law applied to a fluid element of mass $dm = \rho d^3 r$ (where ρ is the fluid's density and $d^3 r$ a volume element) becomes the Navier-Stokes equation, here in integral form after summing over fluid elements,

$$\oint_S T^{ij} n^j dA + \int_V \rho g^i d^3 r = \int_V \rho \frac{dv^i}{dt} d^3 r \tag{2.110}$$

where a fluid volume V is enclosed by surface S and g^i denotes a component of the local gravitational field. Write the stress tensor for a viscous fluid sliding down a plane inclined at the angle α below the horizontal.

2.15 Show that the electric susceptibility tensor χ^{ij} is symmetric. Background: Gauss's law for the electric field says

$$\nabla \cdot \mathbf{E} = \frac{\rho}{\epsilon_o}, \tag{2.111}$$

where ρ is the density of electric charge due to all sources, which includes unbound "free charges" of density ρ_f, and charges due to electric polarization for which $\rho_p = -\nabla \cdot \mathbf{P}$. Writing $\rho = \rho_f + \rho_p$, Gauss's law becomes

$$\nabla \cdot \mathbf{D} = \rho_f \tag{2.112}$$

where, for isotropic dielectrics, $\mathbf{D} \equiv \epsilon_o(1+\chi)\mathbf{E}$. The quantity $1+\chi$ is the dielectric constant κ. For anisotropic dielectrics, the dielectric constant gets replaced with the dielectric tensor

$$D^i = \epsilon_o \kappa^{ij} E^j, \tag{2.113}$$

where $\kappa^{ij} = \delta^{ij} + \chi^{ij}$. The energy density stored in the electrostatic field (i.e., the work necessary to separate the charges and create the fields) equals $\frac{1}{2}\mathbf{D}\cdot\mathbf{E}$. Show from these considerations that $\kappa^{ij} = \kappa^{ji}$ and thus $\chi^{ij} = \chi^{ji}$ (see Panofsky and Phillips, p. 99).

2.16 We have used extensively a simple orthogonal rotation about the z axis through the angle θ, with rotation matrix

$$\Lambda(\theta) = \begin{pmatrix} \cos\theta & \sin\theta & 0 \\ -\sin\theta & \cos\theta & 0 \\ 0 & 0 & 1 \end{pmatrix}. \tag{2.114}$$

Of course, most rotations are more complicated than this. With three-dimensional Cartesian systems, a general rotation can be produced in a succession of three one-axis rotations (each counterclockwise), as follows:
(1) Rotate the original xyz about its z-axis through the angle α to generate the new coordinate axes $x'y'z'$ (where $z'=z$). Let this rotation matrix be denoted $\Lambda_1(\alpha)$.
(2) Next, rotate the $x'y'z'$ system through the angle β about its x' axis to give the new system $x''y''z''$, described by the rotation matrix $\Lambda_2(\beta)$.
(3) Finally, rotate the $x''y''z''$ system through the angle γ about its z'' axis, described by rotation matrix $\Lambda_3(\gamma)$, to give the final set of axes XYZ.
The final result, which describes how the xyz system could get carried into the XYZ system with one rotation matrix, is computed from

$$\Lambda(\alpha,\beta,\gamma) = \Lambda_3(\gamma)\Lambda_2(\beta)\Lambda_1(\alpha). \tag{2.115}$$

(a) Show the final result to be

$$\begin{pmatrix} (\cos\alpha\cos\gamma - \sin\alpha\cos\beta\sin\gamma) & (\sin\alpha\cos\gamma + \cos\alpha\cos\beta\sin\gamma) & (\sin\beta\sin\gamma) \\ (-\cos\alpha\sin\gamma - \sin\alpha\cos\beta\cos\gamma) & (-\sin\alpha\sin\gamma + \cos\alpha\cos\beta\cos\gamma) & (\sin\beta\cos\gamma) \\ (\sin\alpha\sin\beta) & (-\cos\alpha\sin\beta) & (\cos\beta) \end{pmatrix}.$$

The angles α, β, and γ are called "Euler angles" (see, e.g., Marion and Thornton, p. 440).
(b) Show that $\Lambda^\dagger = \Lambda^{-1}$.
(c) Show that, for infinitesimal rotation angles, any orthogonal transformation is approximated by the identity matrix plus an antisymmetric matrix, to first order in angles.

2.17 From the perspective of molecular physics, here we examine a model that enables one to calculate the electric susceptability tensor components χ^{ij}.

If you have encountered time-independent, nondegenerate, stationary state perturbation theory in quantum mechanics, please read on (see Merzbacher, p. 422).

Let N denote the number of molecules per volume, let $\mathbf{E}$ be an applied electric field, denote the average molecular electric dipole moment as $\langle \mathbf{p} \rangle$, and define α^{ij} as a component of the "molecular polarizability" tensor for one molecule, defined by $p^i = \alpha^{ij}E^j$. Thus $P^i = Np^i = N\alpha^{ij}E^j$. But, in addition, $P^i = \epsilon_o \chi^{ij} E^j$ defines χ^{ij}, so that $\chi^{ij} = \frac{N}{\epsilon_o}\alpha^{ij}$. The conceptual difference between α^{ij} and χ^{ij} is that one speaks of the polarizability of the chloroprene molecule (microscopic), but the susceptability of the synthetic rubber material neoprene (macroscopic).

Model the molecule as a 1-electron atom (or any atom with only one valence electron outside a closed shell). The electron carries electric charge $-e$. Let ψ denote the electron's de Broglie wave function. The probability that the electron will be found in volume element dV is $\psi^*\psi dV = |\psi|^2 dV$, where $*$ denotes complex conjugate. The molecular average electric dipole moment will be

$$\langle \mathbf{p} \rangle = -e \int \psi^* \mathbf{r} \psi \, dV. \tag{2.116}$$

Now turn on a small perturbing interaction $U \equiv -\mathbf{p} \cdot \mathbf{E}$. The wave function adjusts from that of the unperturbed atom, and the adjustment shows up as an additive correction to each of the quantized stationary state wave functions, according to

$$\psi_n \approx \psi_n^{(0)} + \psi_n^{(1)}, \tag{2.117}$$

where $\psi_n^{(0)}$ denotes the unperturbed stationary state wave functions for state n that has unperturbed energy $E_n^{(0)}$, and $\psi_n^{(1)}$ denotes its first-order correction caused by the perturbation. This correction, according to perturbation theory, is computed by

$$\psi_n^{(1)} = \sum_{k \neq n} \frac{U_{nk}}{E_n^{(0)} - E_k^{(0)}} \psi_k^{(0)}, \tag{2.118}$$

where the matrix element

$$U_{nk} \equiv \int \psi_n^{*(0)} U \psi_k^{(0)} dV \equiv \langle n|U|k \rangle. \tag{2.119}$$

(a) Show that, to first order in the perturbation, the probability density is

$$\psi^* \psi \approx |\psi_n^{(0)}|^2 + \psi_n^{*(0)} \psi_n^{(1)} + \psi_n^{*(1)} \psi_n^{(0)}. \tag{2.120}$$

(b) In our problem the perturbation is the interaction of the molecule's dipole moment with the $\mathbf{E}$ field, $U = -\mathbf{p} \cdot \mathbf{E}$. Show that, in dyadic notation (recall Ex. 2.11),

$$\begin{aligned}
\langle \mathbf{p} \rangle &= -e \int \psi^* \ \mathbf{r} \ \psi \ dV \\
&= -e\langle n|\mathbf{r}|n \rangle - e^2 \sum_{k \neq n} \frac{\langle n|\mathbf{r}|k \rangle \langle k|\mathbf{r}|n \rangle + \langle k|\mathbf{r}|n \rangle \langle n|\mathbf{r}|k \rangle}{E_n^{(0)} - E_k^{(0)}} \cdot \mathbf{E} \\
&\equiv \langle \mathbf{p}^{(0)} \rangle - \mathbf{p}^{(1)} \cdot \mathbf{E},
\end{aligned}$$

where

$$\langle n|\mathbf{r}|k\rangle \equiv \int \psi_n^{*(0)}\ \mathbf{r}\ \psi_k^{(0)}\ dV. \tag{2.121}$$

The first term, $\langle \mathbf{p}^{(0)} \rangle \equiv \langle n|(-e\mathbf{r})|n\rangle$, denotes the unperturbed molecule's original electric dipole moment due to its own internal dynamics. The second term, $\mathbf{p}^{(1)}$, denotes its dipole moment induced by the presence of the applied $\mathbf{E}$, for which $p^{i(1)} = \alpha^{ij} E^j$.

(c) Write the final expression for the susceptibility tensor χ^{ij} using this model. Is χ^{ij} a product of displacements that will transform as a second-order tensor?

Chapter 3

The Metric Tensor

3.1 The Distinction between Distance and Coordinate Displacement

The metric tensor is so central to our subject that it deserves a chapter of its own. Some authors call it the "fundamental tensor."

It all begins with displacement. Let us write the displacement vector $d\mathbf{r}$, measured in units of length, in three-dimensional Eucildean space, using three different coordinate systems (see Appendix A). In rectangular coordinates,

$$d\mathbf{r} = (dx)\hat{\mathbf{i}} + (dy)\hat{\mathbf{j}} + (dz)\hat{\mathbf{k}}; \tag{3.1}$$

in cylindrical coordinates,

$$d\mathbf{r} = (d\rho)\hat{\boldsymbol{\rho}} + (\rho d\varphi)\hat{\boldsymbol{\varphi}} + (dz)\hat{\mathbf{z}}; \tag{3.2}$$

and in spherical coordinates,

$$d\mathbf{r} = (dr)\hat{\mathbf{r}} + (rd\theta)\hat{\boldsymbol{\theta}} + (r\sin\theta d\varphi)\hat{\boldsymbol{\varphi}}. \tag{3.3}$$

It will be noticed that a *coordinate* displacement is not always a *distance*. In rectangular coordinates, $|dx|$ equals a length, but in cylindrical coordinates, the differential $d\varphi$ needs the radius ρ to produce the distance $\rho|d\varphi|$. The metric tensor components g_{ij}, introduced in this chapter (note the *sub*scripts–stay tuned!), convert coordinate differentials into distance increments.

Displacements can be positive, negative, or zero, but displacements squared (assuming real numbers) are nonnegative. In Euclidean space the infinitesimal distance squared between two points, $(ds)^2 \equiv d\mathbf{r} \cdot d\mathbf{r}$, respects the theorem of Pythagoras. Expressed in rectangular coordinates, the distance between (x, y, z)

and $(x + dx, y + dy, z + dz)$ is given by the coordinate differentials themselves, because they already measure length:

$$(ds)^2 = (dx)^2 + (dy)^2 + (dz)^2. \tag{3.4}$$

Let me pause for a moment: To reduce notational clutter coming from an avalanche of parentheses, $(dx)^2$ (for example) will be written dx^2. Distinctions between the square of the differential, the differential of the square, and superscripts on coordinate displacements (e.g., $dx^2 = dy$) should be clear from the context. When they are not, parentheses will be reinstated.

Back to work: The infinitesimal distance squared, expressed in cylindrical coordinates (ρ, φ, z), reads

$$ds^2 = d\rho^2 + \rho^2 d\varphi^2 + dz^2. \tag{3.5}$$

Here $d\rho$ and dz carry dimensions of length, but $d\varphi$ denotes a dimensionless change in radians.

In spherical coordinates (r, θ, φ) the infinitesimal length appears as

$$ds^2 = dr^2 + r^2 d\theta^2 + r^2 \sin^2\theta d\varphi^2. \tag{3.6}$$

Here two of the three coordinate differentials, $d\theta$ and $d\varphi$, are not lengths, so r and $r\sin\theta$ are required to convert coordinate increments to distances.

Each of these expressions for the infinitesimal length squared, ds^2, can be subsumed into a generic expression by introducing the symmetric "metric tensor" with components g_{ij} (recall the summation convention):

$$ds^2 = g_{ij} dx^i dx^j. \tag{3.7}$$

If all the $g_{\mu\nu}$ are nonnegative, the geometry of the space is said to be "Riemannian," after Georg G. B. Riemann (1826-1866), who will be a towering figure in our subject. If some of the $g_{\mu\nu}$ are negative, the geometry is said to be pseudo-Riemannian. We shall have to deal with both Riemannian and pseudo-Riemannian spaces.

In rectangular coordinates, $dx^1 = dx$, $dx^2 = dy$, $dx^3 = dz$, and $g_{ij} = \delta_{ij}$ (note that the Kronecker delta is here a special instance of the metric tensor and therefore carries subscripts). In cylindrical coordinates, $dx^1 = d\rho$, $dx^2 = d\varphi$, $dx^3 = dz$, with

$$g_{ij} = \begin{pmatrix} 1 & 0 & 0 \\ 0 & \rho^2 & 0 \\ 0 & 0 & 1 \end{pmatrix}. \tag{3.8}$$

In spherical coordinates, $dx^1 = dr$, $dx^2 = d\theta$, $dx^3 = d\varphi$, and

$$g_{ij} = \begin{pmatrix} 1 & 0 & 0 \\ 0 & r^2 & 0 \\ 0 & 0 & r^2\sin^2\theta \end{pmatrix}. \tag{3.9}$$

Euclidean spaces need not be confined to two or three dimensions. A four-dimensional Euclidean space can be envisioned, and its logic is self-consistent.

The distance squared between the points with coordinates (x, y, z, w) and $(x + dx, y + dy, z + dz, w + dw)$ is

$$ds^2 = dx^2 + dy^2 + dz^2 + dw^2 = g_{\mu\nu} dx^\mu dx^\nu, \tag{3.10}$$

where $g_{\mu\nu} = \delta_{\mu\nu}$, represented with a 4×4 matrix. This is not far-fetched at all. For instance, for a particle in motion, one could think of its position (x, y) and its momentum (p^x, p^y) not as two separate vectors in two-dimensional space, but as one vector with the components $(x^1, x^2, x^3, x^4) = (x, y, p^x, p^y)$ in four-dimensional space. This so-called phase space gets heavy use in statistical mechanics.

As suggested above, geometries for which all the g_{ij} are ≥ 0, so that $ds^2 \geq 0$, are called "Riemannian geometries." Spaces for which some of the g_{ij} may be negative, so that ds^2 may be positive, negative, or zero, are called "pseudo-Riemannian" spaces.

The proof that the g_{ij} are components of a second-order tensor will be postponed until later in this chapter, after we have further discussed the reasons for the distinction between superscripts and subscripts. Then we can be assured that the proof holds in pseudo-Rimennian as well as in Riemannian geometries.

3.2 Relative Motion

Henceforth space by itself, and time by itself, are doomed to fade away into mere shadows, and only a kind of union of the two will preserve an independent reality.
–Hermann Minkowski (1908)

Coordinate transformations can include reference frames in relative motion, with time as a coordinate. Transformations to a second reference frame that moves relative to the first one are called "boosts." The set of reference frames that have velocities but no accelerations relative to one another are the inertial frames. Newton's first law is essentially the statement that, for physics to appear as simple as possible, we should do physics in inertial frames. Of course, reality also happens in accelerated frames, and physicists have to consider the relativity of accelerated frames, which modifies $\mathbf{F} = m\mathbf{a}$. But for now let us consider only boosts between inertial frames.

For a vivid mental picture of two inertial frames, imagine the Lab Frame and a Coasting Rocket Frame (see Taylor and Wheeler, *Spacetime Physics*). Observers in the Lab Frame mark events with space and time coordinates (t, x, y, z), and Coasting Rocket observers use coordinates (t', x', y', z'). In the simplest scenario of their relative motion, observers in the Lab Frame see the Coasting

Rocket Frame moving with uniform velocity $\mathbf{v}_r$ in their $+x$-direction. For simplicity let the x and x' axes be parallel, likewise the y and y', z and z' axes. It is also assumed that, within each frame, a set of clocks have been previously synchronized, and for the event where the origins of both frames instantaneously coincide the clocks record $t = 0$ and $t' = 0$. Let us call these circumstances a "simple boost." We now consider a simple boost in Newtonian relativity, followed by a simple boost in the special theory of relativity.

In Newtonian relativity, it is postulated (1) that the laws of mechanics are covariant between all inertial frames, and (2) that length and time intervals between two events are separately invariant. Newton made the latter assumptions explicit. In *The Principia* (1687) he wrote, "Absolute, true, and mathematical time, of itself, and from its own nature flows equably without regard to anything external.... Absolute space, in its own nature, without regard to anything external, remains always similar and immovable." (Motte translation).

An event that has time and space coordinates (t, x, y, z) in the Lab Frame has coordinates (t', x', y', z') in the Coasting Rocket Frame. Under Newtonian assumptions, in a simple boost the two sets of coordinates are related by the "Galilean transformation":

$$\begin{aligned} t' &= t \\ x' &= x - v_r t \\ y' &= y \\ z' &= z. \end{aligned}$$

Consequences of the Galilean transformation include the relativity of velocity. When a particle moves parallel to the x-axis with velocity $v = dx/dt$ in the Lab frame, its velocity along the x'-axis in the Rocket Frame is

$$\begin{aligned} v' &= \frac{dx'}{dt'} \\ &= \frac{dx - v_r dt}{dt} \\ &= v - v_r. \end{aligned}$$

Significantly, if that "particle" happens to be a beam of light moving with speed c through the Lab Frame, then Newtonian relativity predicts that $c' \neq c$. Notice that since $v_r = const$, it follows that $a' = dv'/dt' = dv/dt = a$, that is, acceleration is invariant among inertial frames (the very definition of inertial frames).

In contrast, the special theory of relativity postulates that, among all inertial frames, (1) all the laws of physics—electrodynamics as well as mechanics—are covariant, and (2) the speed of light in vacuum is invariant. Einstein made these postulates explicit: "Examples of this sort...lead to the conjecture that not only the phenomena of mechanics but also those of electrodynamics and optics will be valid for all coordinate systems in which the equations of mechanics hold.... We shall raise this conjecture...to the status of a postulate and shall also introduce another postulate, which is only seemingly incompatible with it, namely that light always propagates in empty space with a definite velocity V that is independent of the state of motion of the emitting body" (1905, Statchel translation).

As logical consequences of these two postulates, one derives the "time dilation" and "length contraction" formulas, to find that $\Delta t \neq \Delta t'$ and $\Delta s \neq \Delta s'$. In particular, if a clock in the Lab Frame reads time interval Δt between two events that occur at the same place in the lab (e.g., the start and end of class, measured by the classroom clock), then identical clocks staked out in the Coasting Rocket Frame, moving with speed v_r relative to the Lab Frame, measure time interval $\Delta t'$ for those same two events (for this observer the start and end of class occur in different places, because this observer sees the classroom zoom by). As shown in introductions to special relativity, the two times are related by

$$\Delta t' = \gamma_r \Delta t, \tag{3.11}$$

where

$$\gamma_r \equiv \frac{1}{\sqrt{1 - \left(\frac{v_r}{c}\right)^2}} \tag{3.12}$$

and c denotes the speed of light in vacuum (Einstein's "V"). If the rocket has length L' as measured by someone aboard it, the rocket's length L as it zooms by in the Lab Frame is

$$L = \frac{L'}{\gamma_r}. \tag{3.13}$$

Although $\Delta t \neq \Delta t'$ and $L \neq L'$, a kind of "spacetime distance" *is* invariant between the frames. For a beam of light, the second postulate requires that $c = ds'/dt' = ds/dt$ so that, for infinitesimal displacements between events connected by the emission and reception of the same light signal,

$$c^2 dt^2 - ds^2 = 0 = c^2 dt'^2 - ds'^2. \tag{3.14}$$

The time dilation and length contraction results further illustrate that, for *any* pair of events, even those not connected by the emission and reception of the same light signal, the so-called spacetime interval is invariant:

$$c^2 dt^2 - ds^2 = c^2 dt'^2 - ds'^2, \tag{3.15}$$

even when it is not equal to zero. The invariance of the spacetime interval serves as an alternative postulate for special relativity, equivalent to the postulate about the invariance of the speed of light.

In the reference frame where the two events occur at the same location ($ds = 0$ but $dt \neq 0$), that particular frame's time interval is called the "proper time" between those events and dignified with the notation $d\tau$. Because this spacetime interval, and thus the numerical value of proper time, is a scalar under boosts, observers in any inertial reference frame can deduce the value of proper time from their metersticks and clocks, even if their frame's clocks do not measure it directly, because of

$$c^2dt^2 - ds^2 = c^2d\tau^2 = c^2dt'^2 - ds'^2. \tag{3.16}$$

If the two events whose time and space separations being measured are the emission and reception of the same flash of light, then the *proper* time between those events equals zero ("lightlike" events); they are separated by the equal amounts of time and space. If $d\tau^2 > 0$, then more time than space exists between the two events ("timelike events"), which means that one event can causally influence the other, because a signal traveling slower than light could connect them. If $d\tau^2 < 0$, then there is more space than time between the events ("spacelike" events), so the *events* (distinct from the *places*) cannot communicate or causally influence one another; not even light can go from one event to the other fast enough.

No generality is lost by considering infinitesimally close events, because for events separated by finite intervals we merely integrate along the path connecting events a and b:

$$\Delta\tau = \int_a^b d\tau. \tag{3.17}$$

For two coordinate systems related by a simple boost, the transformation that preserves the invariance of the spacetime interval, and the invariance of the speed of light, is the simplest version of a "Lorentz transformation":

$$t' = \gamma_r(t - \frac{v_r x}{c^2}) \tag{3.18}$$

$$x' = \gamma_r(x - v_r t) \tag{3.19}$$

$$y' = y \tag{3.20}$$

$$z' = z. \tag{3.21}$$

To make these expressions look cleaner, first move the c's around to write

$$ct' = \gamma_r(ct - x\frac{v_r}{c}) \tag{3.22}$$

$$x' = \gamma_r(x - ct\frac{v_r}{c}). \tag{3.23}$$

Next, absorb c into t, so that henceforth t denotes time measured in meters (in other words, t now denotes the quantity cT with T measured in seconds; thus, if $T = 1$ s, then $t = 3 \times 10^8$m). Also, absorb $1/c$ into the velocities, so that speeds are now expressed as a dimensionless fraction of the speed of light (so that v now means V/c, and if $V = 1.5 \times 10^8$ m/s, then $v = \frac{1}{2}$). Now the simplest Lorentz transformation from the Lab Frame coordinates (t, x, y, z) to the Coasting Rocket Frame coordinates (t', x', y', z') takes the symmetric form

$$t' = \gamma_r(t - v_r x) \tag{3.24}$$

$$x' = \gamma_r(x - v_r t) \tag{3.25}$$

along with the identity transformations $y' = y$ and $z' = z$. Notice that the limit $v_r \ll 1$ (or $v_r \ll c$ in conventional units), the Lorentz transformation reduces to the Galilean transformation.

As an aside, an especially elegant way to parameterize velocities arises by introducing the "rapidity" ϵ_r, defined as $v_r \equiv \tanh \epsilon_r$. For the transformations of Eqs. (3.24)-(3.25) the Lorentz transformation equations are analogous in *appearance* to those of a rotation of axes (but quite different in *meaning*, featuring hyperbolic, not circular, trig functions):

$$t' = t \cosh \epsilon_r - x \sinh \epsilon_r \tag{3.26}$$

$$x' = -t \sinh \epsilon_r + x \cosh \epsilon_r \tag{3.27}$$

$$y' = y \tag{3.28}$$

$$z' = z. \tag{3.29}$$

The nonvanishing partial derivative transformation coefficients are

$$\frac{\partial t'}{\partial t} = \gamma_r = \cosh \epsilon_r \tag{3.30}$$

$$\frac{\partial t'}{\partial x} = -v_r \gamma_r = -\sinh \epsilon_r \tag{3.31}$$

$$\frac{\partial x'}{\partial t} = -v_r \gamma = -\sinh \epsilon_r \tag{3.32}$$

$$\frac{\partial x'}{\partial x} = \gamma_r = \cosh \epsilon_r \tag{3.33}$$

$$\frac{\partial y'}{\partial y} = 1 \tag{3.34}$$

$$\frac{\partial z'}{\partial z} = 1. \tag{3.35}$$

Returning to the main task at hand, I said all of that to say this: Consider two events that are nearby in space and time, where the coordinate displacements between those two events are denoted $dx^0 = dt$, $dx^1 = dx$, $dx^2 = dy$, and $dx^3 = dz$. A set of these four coordinate displacements forms a four-component vector in spacetime, a so-called 4-vector,

$$dx^\mu = (dt, dx, dy, dz) = (dt, d\mathbf{r}). \tag{3.36}$$

The spacetime interval for the proper time squared between those two events can be written with a metric tensor:

$$\begin{aligned} d\tau^2 &= dt^2 - dx^2 - dy^2 - dz^2 \\ &= g_{\mu\nu} dx^\mu dx^\nu, \end{aligned}$$

where

$$g_{\mu\nu} = \begin{pmatrix} 1 & 0 & 0 & 0 \\ 0 & -1 & 0 & 0 \\ 0 & 0 & -1 & 0 \\ 0 & 0 & 0 & -1 \end{pmatrix}. \tag{3.37}$$

This particular metric tensor of Eq. (3.37), which uses Cartesian coordinates for the space sector, will sometimes be denoted $\eta_{\mu\nu} = \pm\delta_{\mu\nu}$ when it needs to be distinguished from other metric tensors. Whatever the coordinates, spacetimes for which

$$d\tau^2 = dt^2 - d\mathbf{r} \cdot d\mathbf{r} \tag{3.38}$$

are called Minkowskian or "flat" spacetimes.

Let a proton emerge from a cyclotron in the Lab Frame. Its velocity in three-dimensional space, as measured by metersticks and clocks within the Lab Frame, is the ratio $d\mathbf{r}/dt$. This notion of what we mean by velocity can be extended so that we speak of a velocity through four-dimensional spacetime. Differentiate the four spacetime coordinates with respect to invariant *proper* time. Why with respect to proper time? For any pair of events, the Lab Frame and all Coasting Rocket Frames agree on the value of the proper time interval, whether or not each frame's clocks measure it directly. Accordingly, the proton's velocity through spacetime (not merely through space), with four components u^μ, is defined according to

$$u^\mu \equiv \frac{dx^\mu}{d\tau}. \tag{3.39}$$

But $x^\mu = x^\mu(t)$, requiring the chain rule:

$$\begin{aligned} u^\mu &= \frac{dx^\mu}{dt}\frac{dt}{d\tau} \\ &= v^\mu \frac{dt}{d\tau}, \end{aligned}$$

where we recognize a component of ordinary velocity $v^\mu = dx^\mu/dt$ in the Lab Frame. But what is $dt/d\tau$? From

$$\begin{aligned} d\tau^2 &= dt^2 - (d\mathbf{r}) \cdot (d\mathbf{r}) \\ &= dt^2 \left(1 - v^2\right) \end{aligned}$$

we obtain

$$\frac{dt}{d\tau} = \frac{1}{\sqrt{1 - v^2}} \equiv \gamma. \tag{3.40}$$

(Notice carefully the distinction between v_r and γ_r that apply to boosts *between* frames on the one hand and, on the other hand, γ and v for the motion of a particle *within* a reference frame.)

These 4-velocity components can be gathered up into another spacetime 4-vector,

$$u^\mu = (u^0, u^1, u^2, u^3) = \gamma(1, \mathbf{v}). \tag{3.41}$$

Notice that, upon dividing $d\tau^2 = g_{\mu\nu}dx^\mu dx^\nu$ by $d\tau$,

$$1 = g_{\mu\nu}u^\mu u^\nu. \tag{3.42}$$

Just as Newtonian momentum $\mathbf{p} = m\mathbf{v}$ is a rescaled velocity, in a similar manner relativistic momentum is defined as a rescaled velocity 4-vector,

$$p^\mu \equiv mu^\mu, \tag{3.43}$$

whose components form a 4-vector,

$$p^\mu = m\gamma(1, \mathbf{v}). \tag{3.44}$$

The $\mu = 1, 2, 3$ components of p^μ are straightforward to interpret, because they reduce to the components of Newtonian momentum $m\mathbf{v}$ in the limit of small velocities. The interpretation of p^0 requires more examination. In a binomial expansion of γ, no terms linear in v occur, but we do find that

$$p^0 = m\gamma = m + \frac{1}{2}mv^2 + \cdots, \tag{3.45}$$

or, in conventional units,

$$p^0 = mc^2 + \frac{1}{2}mv^2 + \cdots \tag{3.46}$$

which contains the Newtonian kinetic energy to lowest order in velocity. Therefore, p^0 can be interpreted as a non-interacting particle's energy E, kinetic plus mass, $p^0 = E = m\gamma = K + m$. The momentum 4-vector components can therefore be identified as

$$p^\mu = m\gamma(1, \mathbf{v}) = (E, \mathbf{p}). \tag{3.47}$$

Since these are the components of vectors in spacetime, their square follows from the metric tensor,

$$g_{\mu\nu}p^\mu p^\nu = E^2 - \mathbf{p} \cdot \mathbf{p} = m^2, \tag{3.48}$$

which also follows by multiplying Eq. (3.42) by m^2. In conventional units, $E = mc^2\gamma$, $\mathbf{p} = m\mathbf{v}\gamma$, and

$$E^2 - (pc)^2 = (mc^2)^2. \tag{3.49}$$

The $(+---)$ "signature" of the signs along the diagonal of the special relativity metric tensor identifies Minkowski spacetime as a pseudo-Riemannian geometry. Some authors uses the opposite signature $(-++-)$; it makes no fundamental difference because the crucial feature of special relativity is that the time and the space sectors of the metric carry *opposite* signs. Whatever the signature, these opposite signs illustrate the necessity for making a distinction that gives rise to the subscript and superscript labels for tensor components, to which we now turn.

3.3 Upper and Lower Indices

In Chapter 2 we used only superscripts to label the components of vectors and other tensors. But when confronted with the formal tensor definition at the outset, we asked, "Why are some indices written as superscripts and others as subscripts?" If tensor calculus were done only in Euclidean spaces, then the distinction between upper and lower indices would be unnecessary. However, in some geometries, the scalar product does not always appear with plus signs exclusively in the sum. We have just seen an example in the Minkowski spacetime of special relativity. When an inertial reference frame measures the time dt and the displacement $d\mathbf{r}$ between two events, the "time squared *minus* space squared" equals the proper time squared between the events, an invariant:

$$d\tau^2 = dt^2 - d\mathbf{r} \cdot d\mathbf{r}. \tag{3.50}$$

In the four dimensions of "curved" spacetime around a spherically symmetric, non-rotating, uncharged star of mass M, general relativity teaches us that the spacetime interval gets modified from the Minkowski metric into the "Schwarzschild metric" (here restoring explicit c's):

$$c^2 d\tau^2 = A(r)c^2 dt^2 - \frac{dr^2}{A(r)} - r^2 d\theta^2 - r^2 \sin^2\theta d\varphi^2, \tag{3.51}$$

where

$$A(r) \equiv 1 - \frac{2GM}{rc^2} \tag{3.52}$$

and G denotes Newton's gravitational constant. Around a Schwarzschild star the metric tensor therefore takes the following form, when using spherical spatial coordinates, with the spatial origin located at the star's center:

$$g_{\mu\nu} = \begin{pmatrix} A(r) & 0 & 0 & 0 \\ 0 & -\frac{1}{A(r)} & 0 & 0 \\ 0 & 0 & -r^2 & 0 \\ 0 & 0 & 0 & -r^2 \sin^2\theta \end{pmatrix}. \tag{3.53}$$

The variety of geometries confronts us with a choice: do we assign the scalar product one definition for Euclidean spaces, a different one in the Minkowskian spacetime of Special Relativity, and yet another definition around a Schwarzschild star? Should we introduce imaginary numbers so the proper time squared in Minkowskian spacetime looks superficially like the four-dimensional Euclidean metric of Eq. (3.10)? We could choose any of these ways, defining the scalar product with the necessary signs and other factors, as needed, on a case-by-case basis. On the other hand, wouldn't it be more elegant—and robust—to maintain a universal definition of the scalar product as a *sum* of products—for Riemannian and pseudo-Riemannian geometries alike—and let the metric tensor carry the distinctive signs and other factors? That way the *meaning* of "scalar product" will be universal among all geometries. The distinctions between the *computation* of their scalar products will be carried in their metric tensors.

The founders of algebra faced a similar choice when articulating the logic of subtraction. Should subtraction be made another operation distinct from addition, or should subtraction be defined in terms of addition? The latter option requires the introduction of a new set of numbers, corresponding one-on-one to the original ones: for every x there exists a $-x$, with the property that $x + (-x) = 0$. Then $y - x$ formally means $y + (-x)$, and subtraction becomes a case of addition.

Similarly, we choose to double the set of vectors while maintaining the formal definition of the scalar product as a sum of products. Whatever the geometry of a space or spacetime we might encounter, vectors will carry indices that henceforth will come in one of two notations: one kind carries upper indices such as p^μ, and the other kind bears lower indices such as p_μ. The index μ stands for 1, 2, or 3 in three-dimensional space, and $\mu = 0, 1, 2, 3$ in four-dimensional spacetime, where $\mu = 0$ denotes the time coordinate. In addition, from now on, in our convention of summing over repeated indices, one of the paired indices must be a superscript, and the other must be a subscript; indeed, in this chapter we have already started doing this with $g_{\mu\nu}dx^\mu dx^\nu$.

Therefore, in any Riemannian or pseudo-Riemannian geoemtry, the scalar product, that defines the invariant distance or the proper time squared (as the case may be), is written according to

$$ds^2 = g_{\mu\nu}dx^\mu dx^\nu, \tag{3.54}$$

where $g_{\mu\nu} = g_{\nu\mu}$.

What rule or convention determines which quantities carry upper indices and which carry lower indices? We originally defined vectors in terms of displacements from point (or event) A to point (or event) B. The displacement vector led to velocity, acceleration, momentum, angular momentum, force, torque vectors, and so on. *The components of "displacement-derived vectors" are denoted with superscripts* and will continue to be called "vectors," because the coordinates themselves are labeled with superscripts, x^μ.

What are the "vectors" whose components carry subscripts? To see them in relation to the superscripted displacement vector components, break $g_{\mu\nu}dx^\mu dx^\nu$ apart into $(g_{\mu\nu}dx^\mu)(dx^\nu)$ and consider the piece $(g_{u\nu}dx^\mu)$. The u gets summed out, leaving behind a displacement dx_ν:

$$dx_\nu \equiv g_{\mu\nu}dx^\mu. \tag{3.55}$$

For instance, setting $\nu = 0$, in the case of special relativity, and with rectangular coordinates for the spatial sector, gives

$$dx_0 = g_{\mu 0}dx^\mu = g_{00}dx^0 + g_{10}dx^1 + g_{20}dx^2 + g_{30}dx^3, \tag{3.56}$$

with only $dx_0 = dx^0 = dt$ left standing as the surviving term. However, for dx_1 we see that $dx_1 = -dx^1 = -dx$, and similarly $dx_2 = -dy$ and $dx_3 = -dz$. One says that the dx_i are "dual" to the dx^i. (In Euclidean spaces, vectors and their duals are redundant; in special relativity the spatial components of a vector and its dual differ by a sign.) Corresponding to the vector with components $dx^\mu = (dt, d\mathbf{r})$, in special relativity we now have its dual vector with its components, which are the $dx_\mu = (dt, -d\mathbf{r})$. In our approach so far, the dx^μ are *defined* by the coordinate system, whereas the dx_μ are *derived* with the help of the $g_{\mu\nu}$.

In traditional terminology one speaks of "contravariant vectors" with components A^μ and "covariant vectors" with components A_μ. It is fashionable today to call the A^μ the components of "vectors" (without the "contravariant" prefix) and the A_μ the components of the "dual vector" or, in other language, the components of a "1-form." In Section 3.5 we also relate the components of vectors and their duals to the components of the familiar "ordinary" vectors. More about the dual vector and 1-form concepts will be presented in Chapters 7 and 8.

The invariant spacetime interval may be written, in Dirac bracket notation, as the scalar product of the vector $|dx\rangle$ and its dual, denoted $\langle dx|$:

$$\begin{aligned} d\tau^2 &= g_{\mu\nu} dx^\mu dx^\nu \\ &= dx_\mu dx^\mu \\ &= \langle dx|dx\rangle. \end{aligned}$$

Similarly, the inner product of any vector and its dual can be written, in any Riemannian or pseudo-Riemannian space, as

$$\begin{aligned} \langle A|B\rangle &= g_{\mu\nu} A^\mu B^\nu \\ &= A_\mu B^\mu \\ &= A^\mu B_\mu, \end{aligned}$$

which is a scalar, as will be demonstrated.

So far "downstairs-indexed" dual vector components are constructed from the "upstairs-indexed" vector components through the use of the metric tensor. But it would be more satisfying if a mathematical object could be found or invented that would "naturally" be represented by one downstairs index, without having to go through $g_{\mu\nu}A^\nu$. Can such a quantity be conceptualized that, on its own, could serve as the prototope for dual vectors, analogous to how displacement serves as the prototype for upstairs-indexed vectors? We find what we seek in the gradient.

By its definition in Cartesian coordinates, where all the coordinates have units of length, the gradient is a vector $\boldsymbol{\nabla}\phi$ whose components $\nabla_i\phi$ are the derivative of a scalar ϕ with respect to a coordinate increment having the dimension of length:

$$\nabla_i \phi = \frac{\partial \phi}{\partial x^i}. \tag{3.57}$$

The gradient so defined is said to be a vector, but one whose coordinate displacement appears in the "denominator" instead of the "numerator"; hence, the index is a subscript. Accordingly, the transformation rule for $\frac{\partial}{\partial x^\mu}$ will be different from the rule for dx^μ.

In rectangular coordinates, the coordinates x^i have the dimensions of length, so differentiating with respect to a length and differentiating with respect to a coordinate are the same. However, in other systems some of the coordinates may be angles. For example, in cylindrical coordinates (ρ, θ, z), evaluating the derivative with respect to θ is not dividing by a length, so the θ component of the gradient in cylindrical coordinates needs a length in the denominator, and we have seen that

$$\nabla_\theta \phi = \frac{1}{\rho}\frac{\partial \phi}{\partial \theta}. \tag{3.58}$$

This is not a definition, but rather a logical consquence of the definition of the familiar Euclidean gradient as the derivative of ϕ with respect to *distance*.

Going beyond Euclidean spaces to generalized coordinates, let ϕ be a scalar function of the coordinates, $\phi = \phi(x^\mu)$. For the purposes of tensor calculus, when we evaluate the gradient of ϕ, we take derivatives with respect to *coordinates* (which may or may not have dimensions of distance), according to

$$\frac{\partial \phi}{\partial x^\mu}. \tag{3.59}$$

As noted above, since the superscript on x^μ in the gradient appears "upstairs in the denominator" as $\partial/\partial x^\mu$, it is notationally consistent and elegant to write a component of the gradient of ϕ using a subscript on the partial derivative symbol, or with a comma and subscript:

$$\frac{\partial \phi}{\partial x^\mu} \equiv \partial_\mu \phi \equiv \phi_{,\mu}. \tag{3.60}$$

To help avoid confusion between commas that denote partial derivatives and commas that are for ordinary punctuation, note that a comma denoting a derivative immediately precedes the subscript or superscript that labels the coordinate with respect to which the partial derivative is evaluated.

Now consider a coordinate transformation from the x^μ to the x'^μ, where each x'^μ is a function of all the x^ν. By the chain rule for partial derivatives we find

$$\frac{\partial \phi}{\partial x'^\mu} = \frac{\partial \phi}{\partial x^\nu}\frac{\partial x^\nu}{\partial x'^\mu}, \tag{3.61}$$

which states the rule for the transformation of the gradient. In subscript notation,

$$(\partial_\mu \phi)' = \frac{\partial x^\nu}{\partial x'^\mu}(\partial_\nu \phi), \tag{3.62}$$

and in a comma notation,

$$\phi'_{,\mu} = \frac{\partial x^\nu}{\partial x'^\mu}\phi_{,\nu}. \tag{3.63}$$

Because the gradient carries lower indices and transforms inversely to a vector, the gradient's transformation rule serves as the prototype for the transformation of all dual vectors. Thus, if A_μ denotes a component of any dual vector, it transforms according to

$$A'_\mu = \frac{\partial x^\nu}{\partial x'^\mu} A_\nu. \tag{3.64}$$

As promised, we can now demonstrate that $\langle A|B\rangle$ is a scalar:

$$\begin{aligned}
\langle A'|B'\rangle &= A'_\mu B'^\mu \\
&= \left(\frac{\partial x^\rho}{\partial x'^\mu} A_\rho\right)\left(\frac{\partial x'^\mu}{\partial x^\sigma} B^\sigma\right) \\
&= \frac{\partial x^\rho}{\partial x'^\mu}\frac{\partial x'^\mu}{\partial x^\sigma} A_\rho B^\sigma \\
&= \delta^\rho{}_\sigma A_\rho B^\sigma \\
&= A_\rho B^\rho \\
&= \langle A|B\rangle.
\end{aligned}$$

Notice in the argument a crucial point, which can be seen as a trivial identity,

$$\delta^\rho{}_\sigma = \frac{\partial x^\rho}{\partial x^\sigma}, \tag{3.65}$$

and since $x^\mu = x^\mu(x'^\nu)$, this is also an instance of the chain rule:

$$\delta^\rho{}_\sigma = \frac{\partial x^\rho}{\partial x'^\mu}\frac{\partial x'^\mu}{\partial x^\sigma}. \tag{3.66}$$

As already noted, the metric tensor of Euclidean space, mapped with Cartesian coordinates, is expressed by the identity matrix, $g_{ij} = \delta_{ij}$. In that geometry a vector with superscript components and its dual with subscript components are redundant. That is why, in Euclidean spaces, there was no need for the distinction between superscripts and subscripts. But making the distinction everywhere from now on, for the sake of consistency, is good practice. Therefore, to smooth out discussions that follow, we will use upper and lower indices in their places, even in Euclidean cases where such distinctions are unnecessary. That is why we initially wrote the components of the inertia tensor in rectangular Euclidean coordinates as

$$I^{ij} = \int dm[\delta^{ij}r^2 - x^i x^j]. \tag{3.67}$$

3.4 Converting between Vectors and Duals

We have seen how a component A_μ of a dual vector can be constructed from its corresponding vector component A^μ via the metric tensor, $A_\mu = g_{\mu\nu}A^\nu$. Of course, we need to be able to go the other way and find the A^μ should we be given the A_μ. This requires the multiplicative inverse of the metric tensor. If g denotes the matrix representation of the metric tensor, then its multiplicative inverse, g^{-1}, has the property that $gg^{-1} = g^{-1}g = \mathbf{1}$, where $\mathbf{1}$ denotes the identity matrix with Kronecker delta matrix elements $\delta_{\mu\nu}$. Let us *define* the components of g^{-1} using upper indices, so that $g^{\mu\nu}$ denotes the $\mu\nu$th component of g^{-1}. Thus, component by component, the $g^{\mu\nu}$ are defined by the system of equations

$$g^{\mu\lambda}g_{\lambda\nu} = \delta^\mu{}_\nu. \tag{3.68}$$

Bearing both upper and lower indices, $\delta^\mu{}_\nu$, which is numerically equal to $\delta_{\mu\nu}$ and $\delta^{\mu\nu}$, offers our first example of a "mixed tensor," in this instance an order-2 tensor that has one contravariant and one covariant index.

For an example of this inversion, suppose we are given this metric tensor in spacetime:

$$g_{\mu\nu} = \begin{pmatrix} A(t,r) & C(t,r) & 0 & 0 \\ C(t,r) & -B(t,r) & 0 & 0 \\ 0 & 0 & -r^2 & 0 \\ 0 & 0 & 0 & -r^2\sin^2\theta \end{pmatrix}. \tag{3.69}$$

The system of equations implied in Eq. (3.68) yields a set of simultaneous equations to be solved for the various components of the inverse metric tensor, and one obtains

$$g^{\mu\nu} = \begin{pmatrix} \frac{B}{D} & \frac{C}{D} & 0 & 0 \\ \frac{C}{D} & -\frac{A}{D} & 0 & 0 \\ 0 & 0 & -\frac{1}{r^2} & 0 \\ 0 & 0 & 0 & -\frac{1}{r^2\sin^2\theta} \end{pmatrix}, \tag{3.70}$$

where $D \equiv AB+C^2$. (Notice that this "ABC metric" contains the Schwarzschild and Minkowskian metrics as special cases.)

With $g^{\mu\nu}$ in hand, one readily converts between vectors and their duals. We already know, from the way that dx_μ was introduced from dx^μ, that $A_\mu = g_{\mu\nu}A^\nu$. Multiply this by $g^{\lambda\mu}$ and then use Eq. (3.68), from which it follows that

$$A^\lambda = g^{\lambda\mu}A_\mu. \tag{3.71}$$

With the same procedures, indices for higher-order tensors can be similarly raised and lowered:

$$\begin{aligned} T^{\mu\nu} &= g^{\mu\rho}g^{\nu\sigma}T_{\rho\sigma}, \\ T^\mu{}_\nu &= g^{\mu\rho}T_{\rho\nu}, \\ T_{\mu\nu} &= g_{\mu\rho}g_{\nu\sigma}T^{\rho\sigma}, \end{aligned}$$

and so on.

Since a component of the gradient (derivative with respect to a superscripted coordinate) is denoted

$$\frac{\partial}{\partial x^\mu} \equiv \partial_\mu, \tag{3.72}$$

a gradient evaluated with respect to a subscripted coordinate can be defined and denoted

$$\frac{\partial}{\partial x_\mu} = \partial^\mu, \tag{3.73}$$

consistent with the raising and lowering operations with the metric tensor,

$$g^{\mu\nu}\partial_\mu = \partial^\nu. \tag{3.74}$$

Specifically, in Minkowski spacetime, the "original" gradient is

$$\partial_\mu = (\frac{\partial}{\partial t}, \mathbf{\nabla}), \tag{3.75}$$

and its dual works out to be

$$\partial^\mu = (\frac{\partial}{\partial t}, -\mathbf{\nabla}). \tag{3.76}$$

By extension of the transformation of the gradient, we can define the tensor transformation law for tensors of higher order which carry lower-index components, such as

$$T'_{\mu\nu\rho} = \frac{\partial x^\alpha}{\partial x'^\mu}\frac{\partial x^\beta}{\partial x'^\nu}\frac{\partial x^\gamma}{\partial x'^\rho}T_{\alpha\beta\gamma}. \tag{3.77}$$

As anticipated above, it is possible to have "mixed tensors" with some indices upstairs and some downstairs. Tensors with a superscripts and b subscripts are said to be of order $a+b$, whose components transform like a product of a vector components and b dual vector components, such as this order 3 = 1+2 tensor

$$T'^\mu{}_{\nu\rho} = \frac{\partial x'^\mu}{\partial x^\alpha}\frac{\partial x^\beta}{\partial x'^\nu}\frac{\partial x^\gamma}{\partial x'\rho}T^\alpha{}_{\beta\gamma}. \tag{3.78}$$

In any Riemannian or pseudo-Riemannian space, we are now in a position to prove that the $g_{\mu\nu}$ are, indeed, the components of a second-order tensor, by showing that $g_{\mu\nu}$ respects the appropriate transformation rule. Let us begin with the fundamental principle that distances in space (for geometry), or proper times in spacetime (for relativity), are invariant. Hence, for two coordinate systems mapping out in the same space interval,

$$ds^2 = ds'^2. \tag{3.79}$$

In terms of the respective coordinate systems and their metric tensors, this says

$$g_{\mu\nu}dx^\mu dx^\nu = g'_{\rho\sigma}dx'^\rho dx'^\sigma. \tag{3.80}$$

We already have the transformation rule for the displacements, and with them we can write Eq. (3.80) as

$$g_{\mu\nu}dx^{\mu}dx^{\nu} = g'_{\rho\sigma}\frac{\partial x'^{\rho}}{\partial x^{\mu}}\frac{\partial x'^{\sigma}}{\partial x^{\nu}}dx^{\mu}dx^{\nu}. \tag{3.81}$$

Transposing, this becomes

$$\left(g_{\mu\nu} - g'_{\rho\sigma}\frac{\partial x'^{\rho}}{\partial x^{\mu}}\frac{\partial x'^{\sigma}}{\partial x^{\nu}}\right)dx^{\mu}dx^{\nu} = 0. \tag{3.82}$$

This must hold *whatever* the displacement, which requires

$$g_{\mu\nu} = \frac{\partial x'^{\rho}}{\partial x^{\mu}}\frac{\partial x'^{\sigma}}{\partial x^{\nu}}g'_{\rho\sigma}, \tag{3.83}$$

the rule for the transformation of a second-order tensor from the primed to the unprimed frame.

When a tensor of order 2 is represented as a square matrix, the left index, whether a subscript or a superscript, denotes the row "r," and the right index denotes the column "c" for that component's location in the matrix:

$$T_r{}^c \tag{3.84}$$

$$T^r{}_c \tag{3.85}$$

$$T_{rc} \tag{3.86}$$

$$T^{rc}. \tag{3.87}$$

Through the metric tensor $g_{\mu\nu}$ or its inverse $g^{\mu\nu}$, these subscripts and superscripts can be raised and lowered, like toggle switches, as we have seen:

$$g_{\sigma\rho}T^{\mu\rho} = T^{\mu}{}_{\sigma} \tag{3.88}$$

and

$$g^{\sigma\rho}T_{\mu\rho} = T_{\mu}{}^{\sigma}. \tag{3.89}$$

With the possibility of raising or lowering the indices, it is good practice when writing tensor components to prevent superscripts from standing directly above subscripts.

3.5 Contravariant, Covariant, and "Ordinary" Vectors

NOTICE: In this section the summation convention will be suspended. Repeated indices are not summed unless the capital sigma is written explicitly. In addition, it will be convenient to use here the terminology of "contravariant" (upper indices) and "covariant" (lower indices) vector components (instead of "vectors" vs. "dual vectors"), wherein the word "vector" will be used in three contexts for the purpose of comparison.

In this section we will clarify how contravariant and covariant vector components are related to the components of the good old "ordinary" vectors we know from introductory physics. Where do those "ordinary vector" components fit into the contravariant/covariant schema?

To be clear, by "ordinary" I mean the coefficients of unit basis vectors in rectangular, spherical, or cylindrical coordinates in Euclidean space. These are the vector components familiar to us from our encounters with them in mechanics and electrodynamics courses on topics where non-Euclidean spaces were not an issue. In this chapter these "ordinary" vector components will be denoted $\tilde{A}_\mu$. Its subscript carries no covariant or contravariant significance; it is merely a label to identify the unit vector for which it serves as a coefficient. For instance, $\tilde{A}_r$ means the coefficient of $\hat{\mathbf{r}}$, and so on. In rectangular Cartesian coordinates (x, y, z), "ordinary" vectors are written

$$\tilde{\mathbf{A}} = \tilde{A}_x\hat{\mathbf{i}} + \tilde{A}_y\hat{\mathbf{j}} + \tilde{A}_z\hat{\mathbf{k}}; \tag{3.90}$$

in cylindrical coordinates (ρ, φ, z),

$$\tilde{\mathbf{A}} = \tilde{A}_\rho\hat{\boldsymbol{\rho}} + \tilde{A}_\varphi\hat{\boldsymbol{\varphi}} + \tilde{A}_z\hat{\mathbf{k}}; \tag{3.91}$$

and in spherical coordinates (r, θ, φ),

$$\tilde{\mathbf{A}} = \tilde{A}_r\hat{\mathbf{r}} + \tilde{A}_\theta\hat{\boldsymbol{\theta}} + \tilde{A}_\varphi\hat{\boldsymbol{\varphi}}. \tag{3.92}$$

The question now addressed is this: how are these "ordinary" components such as $\tilde{A}_\theta$ or $\tilde{A}_\rho$ or $\tilde{A}_x$ related to the contravariant vector components A^μ and to the covariant vector (or dual) components A_μ?

In the familiar Euclidean coordinate systems, where dx^μ denotes a coordinate differential (note the preserved superscript, which still has its contravariant meaning because this is a coordinate displacement) the line element

$$ds^2 = \sum_{\mu\nu} g_{\mu\nu} dx^\mu dx^\nu \tag{3.93}$$

features only diagonal metric tensors with $(+ + +)$ signatures. Thus, each element of the metric tensor can be written as

$$g_{\mu\nu} = h_\mu^2 \delta_{\mu\nu}, \tag{3.94}$$

with no sum over μ. Written out explicitly, such a line element takes the form

$$ds^2 = (h_1)^2(dx^1)^2 + (h_2)^2(dx^2)^2 + (h_3)^2(dx^3)^2. \tag{3.95}$$

A coordinate displacement dx^μ may—or may not—be a length; however, $h_\mu dx^\mu$ (no sum) *does* denote a length. For example, in spherical coordinates,

$$ds^2 = dr^2 + r^2 d\theta^2 + r^2 \sin^2 \theta d\varphi^2, \tag{3.96}$$

so that $h_1 = 1$, $h_2 = r$, and $h_3 = r \sin \theta$. Since contravariant vector components are rescaled coordinate displacements, the $\tilde{A}_\mu$ can be consistently related to contravariant vector components by identifying

$$A^\mu = \frac{\tilde{A}_\mu}{h_\mu} \tag{3.97}$$

(no sum over μ). What reason justifies identifying this as the relation between the contravariant and "ordinary" component of a vector? As with so much in this subject, it goes back to the scalar product. If we want the familiar dot product definition made with "ordinary" vectors, $\tilde{\mathbf{A}} \cdot \tilde{\mathbf{B}}$, to give the same number as the scalar product in generalized coordinates, then we require

$$\begin{aligned} \sum_{\mu\nu} g_{\mu\nu} A^\mu B^\nu &= (h_1)^2 A^1 B^1 + (h_2)^2 A^2 B^2 + (h_3)^2 A^3 B^3 \\ &= \tilde{A}_1 \tilde{B}_1 + \tilde{A}_2 \tilde{B}_2 + \tilde{A}_3 \tilde{B}_3, \end{aligned}$$

which is guaranteed by the relation of Eq. (3.97).

With the contravariant components in hand, the covariant ones follow at once in the usual way,

$$A_\mu = \sum_\mu g_{\mu\nu} A^\nu, \tag{3.98}$$

which along with Eqs. (3.94) and (3.97) gives

$$A_\mu = h_\mu \tilde{A}^\mu \tag{3.99}$$

(no sum).

These relationships are readily illustrated with spherical coordinates in Euclidean space, where the coordinates (x^1, x^2, x^3) mean (r, θ, φ). The contravariant (or displacement-based vector) components are related to the "ordinary" ones by

$$\begin{aligned} A^1 &= \tilde{A}_r \\ A^2 &= \frac{\tilde{A}_\theta}{r} \\ A^3 &= \frac{\tilde{A}_\varphi}{r \sin \theta}, \end{aligned}$$

and the covariant (or gradient-modeled dual vector) components are connected to the "ordinary" ones according to

$$\begin{aligned} A_1 &= \tilde{A}_r \\ A_2 &= r \tilde{A}_\theta \\ A_3 &= r \sin \theta \tilde{A}_\varphi. \end{aligned}$$

As we have seen, the metric tensor's reason for existence is to turn coordinate displacements into distances, and in generalized coordinates it does so by generalizing the job done by the h's that appear in Euclidean space mapped with various coordinate systems.

Covariant, Contravariant, and "Ordinary" Gradients

Let us pursue the relation between the contravariant/covariant gradients ∂^μ and ∂_μ one the one hand and, on the other hand, the "ordinary" gradient ∇. (No tilde is used on the "ordinary" gradient ∇ because, in this book, ∇ *means* the ordinary gradient; some authors, such as Hobson, Efstathiou, and Lasenby use this symbol to mean the "covariant derivative" that will be introduced later.) Given some scalar function $f(x^k)$ for $k = 1, 2, 3$, its "ordinary" gradient ∇f is *defined* in rectangular coordinates (x, y, z) according to derivatives with respect to lengths,

$$\nabla f \equiv \frac{\partial f}{\partial x}\hat{\mathbf{i}} + \frac{\partial f}{\partial y}\hat{\mathbf{j}} + \frac{\partial f}{\partial z}\hat{\mathbf{k}}. \tag{3.100}$$

For instance, the electrostatic field $\mathbf{E}$ is related to the potential ϕ by $\mathbf{E} = -\boldsymbol{\nabla}\phi$, the (negative) derivative with respect to a *length*, and not, in general, the derivative with respect to a *coordinate*, $\partial_i\phi$. The gradient defined as $\boldsymbol{\nabla}\phi$ in Cartesian coordinates must then be transformed to other coordinate grids. When some coordinate x^μ does not carry units of length, the appropriate length is obtained by application of the appropriate h_μ. Thus, the relation between the "ordinary" gradient (derivatives with respect to length, ∇_μ) and the covariant gradient (derivatives with respect to coordinates, ∂_μ) is

$$\nabla_\mu f = \frac{1}{h_\mu}(\partial_\mu f), \tag{3.101}$$

with no sum over μ. For instance, the gradient components in spherical coordinates are

$$\begin{aligned} \nabla_r f &= \frac{1}{h_1}\frac{\partial f}{\partial r} \\ &\equiv \partial_r f \end{aligned}$$

$$\begin{aligned} \nabla_\theta f &= \frac{1}{h_2}\frac{\partial f}{\partial \theta} \\ &\equiv \frac{1}{r}(\partial_\theta f) \end{aligned}$$

and

$$\begin{aligned} \nabla_\varphi f &= \frac{1}{h_3}\frac{\partial f}{\partial \varphi} \\ &\equiv \frac{1}{r\sin\theta}(\partial_\varphi f). \end{aligned}$$

This builds the gradient vector in spherical coordinates:

$$\begin{aligned} \nabla f &= (\nabla_r f)\hat{\mathbf{r}} + (\nabla_\theta f)\hat{\boldsymbol{\theta}} + (\nabla_\varphi f)\hat{\boldsymbol{\varphi}} \\ &= (\partial_r f)\hat{\mathbf{r}} + \left(\frac{1}{r}\partial_\theta f\right)\hat{\boldsymbol{\theta}} + \left(\frac{1}{r\sin\theta}\partial_\varphi f\right)\hat{\boldsymbol{\varphi}}. \end{aligned}$$

Before we examine other derivatives, such as the divergence, the curl, and the Laplacian, and discuss how they are understood in the context of tensors, we first need to examine the notion of "derivative" more carefully. That forms the topic of the next chapter. But algebra comes before calculus, so we first note some points of tensor algebra in the next section.

3.6 Tensor Algebra

Vector algebra operations of scalar multiplication, addition, inner products, and outer products also apply to tensors of arbitrary order and upstairs/downstairs index character. Such algebraic operations for making new tensors from old ones are perhaps best illustrated with examples.

Scalar multiplication: If b is a scalar, then $bR^{\mu\nu\cdots}{}_{\rho\sigma\ldots}$ is a tensor of the same order as $R^{\mu\nu\cdots}{}_{\rho\sigma\ldots}$.

Addition: For tensor components of the same order and the same upstairs/downstairs index composition, component-by-component addition can be defined, such as

$$S^{\mu\nu}{}_\alpha = T^{\mu\nu}{}_\alpha + R^{\mu\nu}{}_\alpha. \tag{3.102}$$

Contraction, an "inner product" (e.g., $\langle A|B\rangle$), reduces the number of indices by two, as in

$$S^\mu{}_\alpha T^{\alpha\rho}{}_\gamma = R^{\mu\rho}{}_\gamma \tag{3.103}$$

and

$$R^{\mu\nu}{}_{\nu\sigma} = R^\mu{}_\sigma. \tag{3.104}$$

An important special case occurs when all the indices of an even-order tensor are contracted, resulting in a scalar:

$$R^{\alpha}{}_{\alpha} = R \tag{3.105}$$

$$S^{\mu\nu}{}_{\mu\nu} = S. \tag{3.106}$$

Proving that such contractions result in scalars can be easily demonstrated; for instance,

$$\begin{aligned} R'^{\mu}{}_{\mu} &= \frac{\partial x'^{\mu}}{\partial x^{\rho}}\frac{\partial x^{\sigma}}{\partial x'^{\mu}} R^{\rho}{}_{\sigma} \\ &= \delta_{\rho}{}^{\sigma} R^{\rho}{}_{\sigma} \\ &= R^{\rho}{}_{\rho}. \end{aligned}$$

Instances of contractions in special relativity include $dx_{\mu}dx^{\mu} = d\tau^2$ and $p_{\mu}p^{\mu} = m^2$.

An outer product (e.g., $|A\rangle\langle B|$) increases the order:

$$R^{\alpha\beta}S_{\rho\sigma} = T^{\alpha\beta}{}_{\rho\sigma}. \tag{3.107}$$

In "ordinary" vector calculus, we know that the gradient of a scalar yields a vector, and the derivative of a vector with respect to a scalar yields another vector. In tensor calculus, in general, does the gradient of a tensor of order N yield a tensor of order $N+1$? Does the derivative of a tensor of order N with respect to a scalar always produce another tensor of order N? These questions are considered in the next chapter. There we run into trouble.

3.7 Tensor Densities Revisited

In Chapter 2 we introduced the inertia and electric quadrupole tensors. As tensors of order 2, their components were proportional to products of two coordinates, which appeared in integrals over distributions of mass or electric charge. For instance, the inertia tensor components are

$$I^{ij} = \int \left[x_k x^k \delta^{ij} - x^i x^j\right] dm. \tag{3.108}$$

We claimed that these components transform as a second-order tensor under a coordinate transformation $x^i \to x'^i$. But the mass increment dm can be written as the product of a mass density ρ and a volume element,

$$dm = \rho dx^1 dx^2 dx^3 \equiv \rho d^3x. \tag{3.109}$$

Therefore, as we asked in Section 2.9, shouldn't we determine the transformation behaviors of the volume element $dx^1dx^2dx^3$ and the density, before declaring the I^{ij} to be components of a second-order tensor? Similar questions should be asked about the electric quadrupole and octupole tensors with their integration measure dq that would be written as a charge density times a volume element.

In those particular cases we got away without considering the transformation properties of ρd^3x because dm and dq are scalars. But those instances do not advance our understanding of the transformation properties of densities and volume elements.

One approach proceeds case-by-case. For instance, in the context of special relativity, the relation between a charge or mass density ρ in the Lab Frame and its density ρ' in the Coasting Rocket Frame follows from the invariance of charge. In the case of the simplest Lorentz transformation between these two frames, the length contraction formula, Eq. (3.13), applies to the relative motion parallel to the x and x' axes. This, together with the experimental observation that charge is invariant, gives the relativity of charge density:

$$\begin{aligned} dq' &= dq \\ \rho'\ dx'dy'dz' &= \rho\ dxdydz \\ \rho'\ \frac{dx}{\gamma_r}dydz &= \rho\ dxdydz, \end{aligned}$$

and thus

$$\rho' = \gamma_r \rho. \tag{3.110}$$

The same conclusion holds for mass density.

Another approach would be to discuss the relativity of volumes and densities for Riemannian or pseudo-Riemannian spaces in general. As a product of three displacements, the volume element in three dimensions d^3x is not a scalar–we might be tempted to suppose that, by itself, it forms a third-order tensor. If mass and electric charge are scalars, then the density ρ cannot be a scalar either, because its product with d^3x somehow results in a scalar, dm or dq. Hence the subject of "tensor densities."

To get a handle on the issue, let us examine the relativity of area in a coordinate transformation. Then we can generalize its result to volumes.

Relativity of Area and the Jacobian

A well-known theorem shows that, under a coordinate transformation $(x, y) \to (x', y')$, an area transforms according to

$$dx'dy' = J\ dxdy, \tag{3.111}$$

where

$$J \equiv \left| \frac{\partial x'}{\partial x} \right|, \tag{3.112}$$

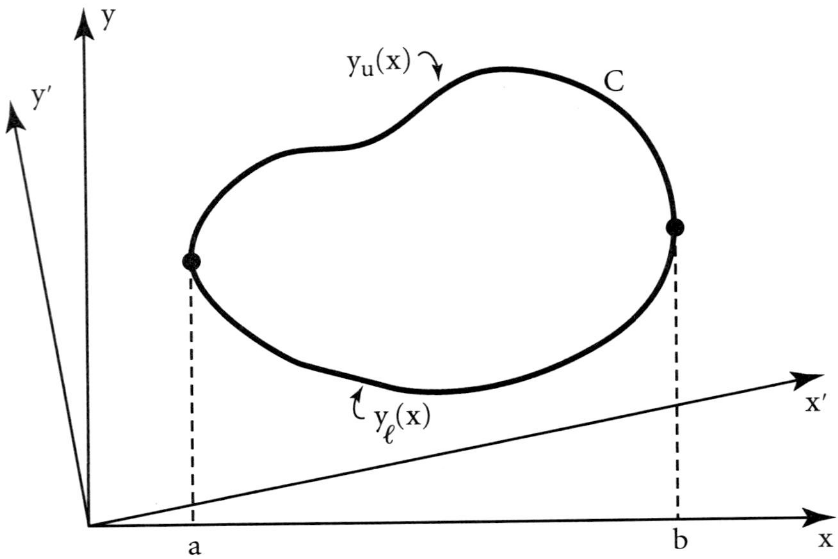

Figure 3.1: *The area and coordinates used to illustrate the relativity of area.*

called the "Jacobian of the x'^μ with respect to the x^μ," is the determinant of the matrix made from the transformation coefficients $\partial x'^\mu/\partial x^\nu$.

To prove the theorem, suppose a closed curve $\mathcal{C}$ forms the boundary of an area A. Lay out this area on a set of axes x and y, and also map the same area with another coordinate system x' and y' (See Fig. 3.1).

Let us calculate area A using the xy axes. Suppose we find the points $x = a$ and $x = b$ where the tangent lines to $\mathcal{C}$ are vertical. Let us call the curve above those points of tangency $y_u(x)$ (u for "upper") and the curve below $y_l(x)$ (l for "lower"). Therefore, the area A enclosed by $\mathcal{C}$ follows from

$$\begin{aligned} A &= \int_a^b y_u dx - \int_a^b y_l dx \\ &= -\int_b^a y_u dx - \int_a^b y_l dx \\ &= -\oint_{\mathcal{C}} y dx. \end{aligned}$$

When calculating the same area using the $x'y'$ axes, by the same reasoning we obtain

$$A' = -\oint_{\mathcal{C}} y' dx'. \tag{3.113}$$

Suppose the $x'y'$ axes are related to the xy axes by the transformation

$$\begin{aligned} x' &= f(x,y) \\ y' &= g(x,y). \end{aligned}$$

When placed in A' of Eq. (3.113), these transformations become

$$\begin{aligned} A' &= -\oint_C g\left[\frac{\partial f}{\partial x}dx + \frac{\partial f}{\partial y}dy\right] \\ &= -\oint_C [E_x dx + E_y dy], \end{aligned}$$

where

$$E_x \equiv g\,\frac{\partial f}{\partial x} \equiv g f_x \tag{3.114}$$

and

$$E_y \equiv g\,\frac{\partial f}{\partial y} \equiv g f_y. \tag{3.115}$$

Now recall Stokes's Theorem,

$$\oint_C \mathbf{E}\cdot d\mathbf{r} = \iint_S (\boldsymbol{\nabla}\times\mathbf{E})\cdot\hat{\mathbf{n}}da, \tag{3.116}$$

where $\mathcal{S}$ denotes the area bounded by closed path $\mathcal{C}$, and $\hat{\mathbf{n}}$ denotes the unit vector normal to the patch of area da, with the direction of $\hat{\mathbf{n}}$ given by a right-hand rule relative to the sense in which the circuit of $\mathcal{C}$ is traversed. For our curve in the xy plane, Stokes's Theorem gives

$$\oint_C [E_x dx + E_y dy] = \iint_S \left[\frac{\partial E_y}{\partial x} - \frac{\partial E_x}{\partial y}\right] dxdy. \tag{3.117}$$

With this, the area A' of Eq. (3.113) may be written

$$\begin{aligned} A' &= -\iint_{A'} \left[\frac{\partial(gf_y)}{\partial x} - \frac{\partial(gf_x)}{\partial y}\right] dxdy \\ &= -\iint_{A'} (g_x f_y - g_y f_x) dxdy \\ &= \iint_{A'} \left[\frac{\partial x'}{\partial x}\frac{\partial y'}{\partial y} - \frac{\partial y'}{\partial x}\frac{\partial x'}{\partial y}\right] dxdy. \end{aligned}$$

The Jacobian J of the transformation from (x,y) to (x',y') is by definition the determinant of the transformation coefficients (the determinant of the Λ of Chs. 1 and 2):

$$J \equiv \begin{vmatrix} \frac{\partial x'}{\partial x} & \frac{\partial x'}{\partial y} \\ \frac{\partial y'}{\partial x} & \frac{\partial y'}{\partial y} \end{vmatrix} \equiv \left|\frac{\partial x'}{\partial x}\right|. \tag{3.118}$$

Therefore,

$$A' = \iint_{A'} J dx dy. \tag{3.119}$$

But, in addition,

$$A' = \iint_{A'} dx' dy'. \tag{3.120}$$

Comparing the two expressions for A' shows that

$$dx'dy' = J dx dy, \tag{3.121}$$

QED.

Had we computed the transformed volume element, we would have obtained

$$dx'dy'dz' = J dx dy dz, \tag{3.122}$$

where J is the determinant of a 3×3 matrix of transformation coefficients (see Boas, p. 171).

The Jacobian can be related to the metric tensor $g_{\mu\nu}$. As a second-order tensor with downstairs indices, the $g_{\mu\nu}$ transform according to

$$g'_{\mu\nu} = \frac{\partial x^\rho}{\partial x'^\mu} \frac{\partial x^\sigma}{\partial x'^\nu} g_{\rho\sigma}. \tag{3.123}$$

Now take the determinant of both sides:

$$|g'| = \left| \frac{\partial x}{\partial x'} \right|^2 |g|. \tag{3.124}$$

The determinant of $g_{\mu\nu}$ requires further comment. In pseudo-Riemannian geometries this determinant can be negative. An immediate example arises in Minkowski spacetime with Cartesian spatial coordinates, for which $\det g_{\mu\nu} = -1$. Therefore, by $|g|$ we mean the absolute value of det $g_{\mu\nu}$; see Question 3.9.

The Jacobian as we introduced it above is the determinant of transformation coefficients from the original unprimed to the new primed coordinates; that is, the Jacobian we called J is the determinant of the coeffiicients $\frac{\partial x'^\mu}{\partial x^\nu}$, which we abbreviated as $|\frac{\partial x'}{\partial x}|$. The Jacobian for the reverse transformation from the "new" back to the "original" coordinates has the form $|\frac{\partial x}{\partial x'}|$. Since the forward-and-reverse transformations in succession yield an identity transformation, it follows that (see Ex. 3.12)

$$\left| \frac{\partial x'}{\partial x} \right| = \left| \frac{\partial x}{\partial x'} \right|^{-1}. \tag{3.125}$$

Therefore, Eq. (3.124) may be written

$$|g'| = \left| \frac{\partial x'}{\partial x} \right|^{-1} \left| \frac{\partial x}{\partial x'} \right| |g|, \tag{3.126}$$

and thus

$$\left|\frac{\partial x'}{\partial x}\right| |g'| = \left|\frac{\partial x}{\partial x'}\right| |g|. \tag{3.127}$$

It will also be seen that, in terms of the Jacobian we called J, $|g'| = J^{-2}|g|$, and hence

$$J = \sqrt{\frac{|g|}{|g'|}}. \tag{3.128}$$

Now Eq. (3.121) may be written in terms of the metric tensor determinants as

$$\sqrt{|g'|}dx'dy' = \sqrt{|g|}dxdy. \tag{3.129}$$

In summary, the Jacobian may prevent an area from being a scalar, but $\sqrt{|g|}dxdy$ makes an invariant area. These results generalize to volumes in Riemannian and pseudo-Riemannian spaces of n dimensions: the volume element transforms as

$$d^n x' = J d^n x, \tag{3.130}$$

with J the determinant of an $n \times n$ matrix. An invariant volume element of n dimensions is

$$\sqrt{|g'|}d^n x' = \sqrt{|g|}d^n x. \tag{3.131}$$

To illustrate, in three-dimensional Euclidean space mapped by Cartesian xyz coordinates, $|g| = 1$ and the same space mapped with spherical coordinates has $|g'| = r^2 \sin\theta$. The product $dxdydz$ is a volume with dimensions of length cubed, but $drd\theta d\varphi$ is *not* a volume in space because not all the coordinates have dimensions of length. However,

$$dxdydz = r^2 \sin\theta \; drd\theta d\varphi, \tag{3.132}$$

which agrees with a coordinate-by-coordinate change of variables from Cartesian to spherical coordinates.

Putting these considerations such as $dq' = dq$ together with their relation to volume elements and charge (or mass) densities, the ratio of volume elements is seen to be equivalent to other expressions:

$$\frac{d^n x'}{d^n x} = \sqrt{\frac{|g|}{|g'|}} = \frac{\rho}{\rho'} = J. \tag{3.133}$$

These imply that the density, like the volume element, does not *quite* transform as a scalar; the factor that keeps it from transforming as a scalar is a power of the Jacobian:

$$\rho' = J^{-1}\rho. \tag{3.134}$$

Conversely, as a consistency check, had we started with Eq. (3.133), the charge and mass could have been *shown* to be scalars, thanks to the Jacobian:

$$\begin{aligned} dq' &= \rho' d^n x' \\ &= (J^{-1}\rho)(J d^n x) \\ &= \rho d^n x \\ &= dq, \end{aligned}$$

and similarly for dm.

Any quantity Q that transforms as

$$Q' = J^W Q \tag{3.135}$$

is said to be a "scalar density of weight W." The square root of the determinant of the metric tensor (or the square root of its absolute value if $\det g_{\mu\nu} < 0$) is a scalar density of weight -1; the volume element is a scalar density of weight $+1$, and a mass or electric charge density is a scalar density of weight -1.

3.8 Discussion Questions and Exercises

Discussion Questions

Q3.1 Since the Earth spins on its axis and revolves around the Sun, and the solar system orbits the Milky Way galactic center (not to mention galactic motions), do *exactly* inertial frames exist in the real world? If not, how are they conceptually meaningful? Discuss whether the following procedure is useful: To see experimentally if your reference frame is inertial, watch a particle on which allegedly no net force acts. If the particle's acceleration is consistent with zero to one part out of 100,000, $|\mathbf{a}| = (0.00000 \pm 0.00001)\text{m/s}^2$, how small would an overlooked nonzero force have to be so that a 0.1 kg particle could be considered to be moving freely?

Q3.2 Explain how the kinetic energy of a particle can be a scalar in Newtonian relativity (under the Galilean transformation), but not in special relativity (under the Lorentz transformation).

Q3.3 Why does $g_{\mu\nu} = g_{\nu\mu}$?

Q3.4 For dealing with the minus signs in the metric of Special Relativity,

$$d\tau^2 = dt^2 - (dx^2 + dy^2 + dz^2) = dt^2 - ds^2, \tag{3.136}$$

we could replace distance ds with an imaginary number ids, to make the Minkowski spacetime interval appear superficially Euclidean. Why isn't this the dominant practice in special relativity? See "Farewell to *ict*" in Misner, Thorne, and Wheeler, p. 51. They mention *ict* instead of *ids* because they discuss a Minkowski metric with the $(- + ++)$ signature.

Q3.5 Suppose the numerical value of the Jacobian is $J = 1$ for some coordinate transformation. Can we then say that volume element is a genuine scalar? In other words, does the *conceptual* distinction between "tensor" and "tensor density" hold even if $J = 1$?

Q3.6 The physicist Paul Dirac once delivered a lecture titled "Methods in Theoretical Physics," which was presented in Trieste, Italy, and later published in the first of the Dirac Memorial Lectures (1988). Dirac said,

> *As relativity was then understood, all relativistic theories had to be expressible in tensor form. On this basis one could not do better than the Klein-Gordon theory* [that describes the evolution of the quantum state for integer-spin particles]. *Most physicists were content with the Klein-Gordon theory as the best possible relativistic theory for an electron, but I was always dissatisfied with the discrepancy between it and general principles, and continually worried over it until I found the solution.*
>
> *Tensors are inadequate and one has to get away from them, introducing two-valued quantities, now called spinors. Those people who were too familiar with tensors were not fitted to get away from them and think up something more general, and I was able to do so only because I was more attached to the general principles of quantum mechanics than to tensors....One should always guard against getting too attached to one particular line of thought.*
>
> *The introduction of spinors provided a relativistic theory in agreement with the general principles of quantum mechanics, and also accounted for the spin of the electron, although this was not the original intention of the work. But then a new problem appeared, that of negative energies.*
>
> - P. A. M. Dirac (in Salam, pp. 135-136)

What is Dirac talking about here? Find out whatever you can about these spinors, and what he means about their being "two-valued" (see, e.g., Dirac, *Principles of Quantum Mechanics*; Feynman, Leighton, and Sands; Aitchison and Hey; Bjorken and Drell; Cartan). I should mention that Dirac did not throw tensors out; rather, he extended relativistic quantum theory to include these spinors as a new family of objects, along with tensors. To manipulate the spin degrees of freedom described by the spinors, Dirac introduced some "hypercomplex numbers" denoted γ^μ and represented as 4×4 matrices. If ψ is one of these spinors (represented as a 4-component column matrix with complex number entries), then $\bar{\psi}$ was a type of adjoint, and the Dirac equation (which offers an alternative to the Klein-Gordon equation in relativistic quantum theory, but applies to spin-$\frac{1}{2}$ particles) was made of "bilinear covariants," which include scalars $\bar{\psi}\psi$, vectors $\bar{\psi}\gamma^\mu\psi$, and second-order tensors $\bar{\psi}\gamma^\mu\gamma^\nu\psi$ (along with pseudoscalars, pseudovectors, and pseudotensors that change sign under spatial inversions, relevant for describing beta decay). The Dirac equation still respects tensor general covariance under a Lorentz transformation, to be consistent with special relativity. But the spinors have a different transformation rule from tensors. By the way, Dirac resolved the "negative energy problem" (viz., an electron at rest apparently had energy $E = -mc^2$) by reinterpreting the minus sign to successfully predict the existence of antimatter.

Q3.7 When showing that

$$dx'dy' = J\ dxdy, \tag{3.137}$$

why didn't we just write

$$dx'^\mu dx'^\nu = \frac{\partial x'^\mu}{\partial x^\rho}\frac{\partial x'^\nu}{\partial x^\sigma}dx^\rho dx^\sigma \tag{3.138}$$

and let it go at that?

Q3.8 In a textbook on relativistic quantum mechanics in Minkowskian spacetime, J. J. Sakurai writes, "Note that we make no distinction between a covariant and a contravariant vector, nor do we define the metric tensor $g_{\mu\nu}$. These complications are absolutely unnecessary in the *special* theory of relativity. (It is regrettable that many textbook writers do not emphasize this elementary point.)" (Sakurai, p. 6). Defend Sakurai's position: How can "these complications" be made unnecessary in special relativity? How would we define the scalar product of two 4-vectors without the metric tensor? Under what circumstances is it advisable to introduce the complexities of the metric tensor, and the covariant and contravariant vectors?

Q3.9 When taking the determinant of the metric tensor during the discussion of the Jacobian and tensor densities, it was necessary to comment that for pseudo-Riemannian spaces (such as the spacetime of special and general relativity), $|g|$ was to be interpreted as the *absolute value* of the determinant of the metric tensor. For example, in Minkowskian spacetime, $\det g = -1$. Weinberg (p. 98) and Peacock (p. 13) define the determinant of the metric tensor, which they denote as g, according to

$$g \equiv -det\ g_{\mu\nu}. \tag{3.139}$$

Is this equivalent to our procedure of handling the determinant of the metric tensor?

Q3.10 Offer an argument that the electric charge is an invariant, without having to resort to a Jacobian. Hint: outside of spacetime arguments, what else do we know about electric charge?

Exercises

3.1 Find the $g^{\mu\nu}$ from the set of equations $g^{\mu\rho}g_{\rho\nu} = \delta^{\mu}{}_{\nu}$ for the Schwarzschild metric.

3.2 Derive the Galilean transformation, starting from the assumption that length and time intervals are separately invariant. Suggestion: consider two specific events, as viewed from the Lab and Coasting Rocket frames.

3.3 (a) Confirm that $\mathbf{F} = m\mathbf{a}$ implies $\mathbf{F}' = m\mathbf{a}'$ under the Galilean transformation.
(b) Confirm that $d\mathbf{r}/dt$ is a tensor of order 1 under a Galilean transformation.

3.4 Consider a simple transformation from the Lab Frame to the Coasting Rocket Frame subject to the postulates of special relativity. Assume $dt^2 - dx^2 = dt'^2 - dx'^2$ and that the transformation from the (t, x) coordinates to the (t', x') coordinates is linear,

$$\begin{aligned} t' &= At + Bx \\ x' &= Ct + Dx, \end{aligned}$$

where A, B, C, and D are constants.
(a) Derive the simplest Lorentz transformation in terms of v_r.
(b) Now reverse the logic and show that the simple Lorentz transformation, taken as given, ensures the invariance of the spacetime interval.

3.5 If observers in the Lab Frame and other observers in the Coasting Rocket Frame watch the same particle moving along their respective x- and x'-directions, derive the relativity of velocity from the simple Lorentz transformation. What do observers in the Coasting Rocket Frame predict for the speed of light c' in the Coasting Rocket Frame if the speed of light is c in the Lab Frame?

3.6 (a) Derive the simple Lorentz transformation for energy and momentum.
(b) Derive from part (a) the Doppler effect for light in vacuum, by using quantum expressions for photon energy and momentum, $E = h\nu$ and $p = h/\lambda$, where h denotes Planck's constant, ν the frequency, and λ the wavelength of a harmonic electromagnetic wave.

3.7 Derive an expression for the transformation of acceleration for a particle that moves through an inertial reference frame, according to the space and time precepts of special relativity. That is, derive in the simple Lorentz transformation an expression for a'^{μ}, where

$$a^{\mu} = \frac{d^2 x^{\mu}}{d\tau^2}. \tag{3.140}$$

3.8 Transform **E** and **B** to **E**$'$ and **B**$'$ (recall Sec. 1.8) in the case where the Lab Frame sees a static infinite line of charge lying along the x axis, carrying uniform charge per length λ. Let the Coasting Rocket Frame move with velocity v_r parallel to the line of charge.

3.9 Suppose that at a particular point P in a particular four-dimensional coordinate system, the metric tensor components have these numerical values:

$$g_{\mu\nu} = \begin{pmatrix} -2 & 0 & 0 & 3 \\ 0 & 1 & 0 & 0 \\ 0 & 0 & 1 & 0 \\ 3 & 0 & 0 & 1 \end{pmatrix}. \tag{3.141}$$

Also consider the vectors that, at P, have the values $A^{\mu} = (1, 0, 4, 2)$ and $B^{\mu} = (1, 5, 0, 1)$. At P, find numerical values for the quantities asked for in parts (a) through (g):
(a) the dual vector components A_{μ} and B_{μ};
(b) the scalar product $A_{\mu}B^{\mu}$;
(c) $g^{\mu\nu}$;
(d) From the dual vector results of part (a), compute the numerical values of the components of the second-order tensor $W_{\mu\nu} = A_{\mu}B_{\nu}$ and $W^{\mu\nu} = A^{\mu}B^{\nu}$;
(e) Calculate $W^{\mu}{}_{\nu}$;
(f) Evaluate $W^{\mu}{}_{\mu}$.
(g) Now consider another tensor evaluated at P,

$$F_{\mu\nu} = \begin{pmatrix} 0 & -1 & -2 & -1 \\ 1 & 0 & -3 & 0 \\ 2 & 3 & 0 & 0 \\ 1 & 0 & 0 & 0 \end{pmatrix}. \tag{3.142}$$

Construct the new tensor components $S_{\mu}{}^{\nu} = F_{\mu\rho}W^{\rho\nu}$.

3.10 Gravitation is often visualized as the "curvature of spacetime." Here we address the question of "What does that mean?" In the Schwarzschild metric of Eq. (3.53), the r-coordinate is *not* the radial distance from the origin; rather, it is defined as follows: Circumnavigate the origin on a spherical shell centered on the origin. Measure the distance traveled in one orbit, the circumference C. Then, by definition, the r-coordinate of that shell is assigned according to

$r \equiv C/2\pi$.
(a) For two concentric spherical shells with r-coordinates r_1 and $r_2 > r_1$, show that the distance between them is given by the integral

$$\int_{r_1}^{r_2} \frac{dr}{1 - \frac{2GM}{rc^2}}, \tag{3.143}$$

and show that it does not equal $r_2 - r_1$.
(b) Suppose $r_1 = 10$ km and $r_2 = 11$ km. What is the difference between the integral and $r_2 - r_1$ when $GM/c^2 = 1.477$ km (the Sun)? When $GM/c^2 = 4 \times 10^9$ km (a black hole)? When $GM/c^2 = 0.44$ cm (the Earth)? This discrepancy between the definite integral and the distance between the two shells as measured by extending a meterstick radially between them is one observable artifact of a "curved space." (See Taylor and Wheeler, *Exploring Black Holes*, Ch. 2.)

3.11 (a) Show by using the tensor transformation rules that $A^{\mu\nu}B_\nu$ is a vector, given that $A^{\mu\nu}$ is an order-2 tensor and B_ν is a dual vector.
(b) Show that $A_{\mu\nu}B^\nu$ is a dual vector.

3.12 In this chapter it was argued informally that

$$\left|\frac{\partial x}{\partial x'}\right| = \left|\frac{\partial x'}{\partial x}\right|^{-1}. \tag{3.144}$$

Show this rigorously. It may be advantageous to make use of the familiar result

$$\frac{\partial x^\rho}{\partial x'^\mu}\frac{\partial x'^\mu}{\partial x^\sigma} = \delta^\rho{}_\sigma. \tag{3.145}$$

3.13 (a) Show that an arbitrary order-2 tensor $T_{\mu\nu}$ may always be written in the form

$$T_{\mu\nu} = A_{\mu\nu} + S_{\mu\nu}, \tag{3.146}$$

where

$$S_{\mu\nu} = S_{\nu\mu} \tag{3.147}$$

and

$$A_{\mu\nu} = -A_{\nu\mu}. \tag{3.148}$$

Under what conditions might this be useful?
(b) Consider an antisymmetric tensor $A_{\mu\nu} = -A_{\nu\mu}$ and a symmetric tensor $S_{\mu\nu} = S_{\nu\mu}$. Show that $A_{\mu\nu}S^{\mu\nu} = 0$.

3.14 Show that the Levi-Civita symbol ε^{ijk} in three-dimensional Euclidean space is a tensor density of order 1 (see Weinberg, Sec. 4.4).

3.15 In Section 3.5 it was demonstrated that the metric tensor is indeed a second-order tensor by showing that

$$g_{\mu\nu} = \frac{\partial x'^{\rho}}{\partial x^{\mu}} \frac{\partial x'^{\sigma}}{\partial x^{\nu}} g'_{\rho\sigma}, \tag{3.149}$$

which is the rule for the transformation of a second-order tensor from the primed to the unprimed frame. From this, or by starting from scratch, demonstrate that, in more conventional arrangement going from unprimed to primed coordinates,

$$g'_{\mu\nu} = \frac{\partial x^{\rho}}{\partial x'^{\mu}} \frac{\partial x^{\sigma}}{\partial x'^{\nu}} g_{\rho\sigma}. \tag{3.150}$$

3.16 Consider the tranformation for a tensor density of weight W, such that

$$D'^{\mu\nu}{}_{\tau} = J^{W} \frac{\partial x'_{\mu}}{\partial x^{\alpha}} \frac{\partial x'^{\nu}}{\partial x^{\beta}} \frac{\partial x^{\gamma}}{\partial x'^{\tau}} D^{\alpha\beta}{}_{\gamma}. \tag{3.151}$$

Would $|g|^{W/2} D^{\mu\nu}{}_{\tau}$ be a tensor? Justify your answer.

3.17 Show that there is no inconsistency in special relativity, for a simple proper Lorentz transformation, in the comparison between

$$dx'dy'dz' = \frac{dx}{\gamma_r} dydz \tag{3.152}$$

and

$$dt'dx'dy'dz' = J\ dtdxdydz. \tag{3.153}$$

3.18 Derive the relations between the ∂^k and ∇_k in three-dimensional Euclidean space, in (a) rectangular and (b) spherical coordinates.

Chapter 4

Derivatives of Tensors

4.1 Signs of Trouble

So many physics definitions are expressed in terms of derivatives! The definitions of velocity and acceleration spring to mind, as does force as the negative gradient of potential energy. So many physics principles are written as differential equations! Newton's second law and Maxwell's equations are among familiar examples. If some definitions and fundamental equations must be written as tensors to guarantee their general covariance under reference frame transformations, then the derivative of a tensor had better be another tensor! The issues at stake can be illustrated with the relativity of velocity and acceleration. Let us begin with the Newtonian relativity of orthogonal and Galilean transformations.

In three-dimensional Euclidean space, a coordinate transforms according to

$$x'^i = \frac{\partial x'^i}{\partial x^j} x^j. \tag{4.1}$$

To derive the transformation of velocity, we must differentiate this with respect to time, a Newtonian invariant. Noting that each new coordinate is a function of all the old coordinates, with the help of the chain rule we write

$$\begin{aligned}
\frac{dx'^i}{dt} &= \frac{d}{dt}\left(\frac{\partial x'^i}{\partial x^j} x^j\right) \\
&= \frac{\partial x'^i}{\partial x^j}\frac{dx^j}{dt} + \frac{\partial^2 x'^i}{\partial x^j \partial x^k}\frac{dx^k}{dt} x^j \\
&= \frac{\partial x'^i}{\partial x^j} v^j + \frac{\partial^2 x'^i}{\partial x^j \partial x^k} v^k x^j,
\end{aligned}$$

where $v^k \equiv dx^k/dt$. The first term on the right-hand side is the condition for velocity to be a first-order tensor. What are we to make of the second term that features the second derivative of a new coordinate with respect to old ones? If it is nonzero, it will spoil the tensor transformation rule for velocity to be a first-order tensor. In the orthogonal-Galilean transformation, the transformations include expressions like $x' = x\cos\theta + y\sin\theta$ for the orthogonal case and $x' = x - v_r t$ for the Galilean transformation. With fixed θ and fixed v_r, the second derivatives all vanish, leaving

$$v'^i = \frac{\partial x'^i}{\partial x^j} v^j. \tag{4.2}$$

Therefore, under orthogonal-Galilean transformations, velocity is a first-order tensor. The order-1 tensor nature of acceleration and other vectors can be tested in a similar way.

So far, so good: at least in the Newtonian relativity of inertial frames, the derivative of a vector with respect to an invariant is another vector. Does the same happy conclusion hold in the spacetime transformations of special and general relativity? Whatever the transformation between coordinate systems in spacetime, consider a coordinate velocity $dx^\mu/d\tau$. Is this the component of an order-1 tensor, the way dx^i/dt was under orthogonal-Galilean transformations? To answer that question, we must investigate how $dx^\mu/d\tau$ behaves under a general spacetime transformation. Since coordinate differentials transform according to

$$dx'^\mu = \frac{\partial x'^\mu}{\partial x^\nu} dx^\nu, \tag{4.3}$$

we must examine

$$\frac{dx'^\mu}{d\tau} = \frac{d}{d\tau}\left(\frac{\partial x'^\mu}{\partial x^\nu} x^\nu\right). \tag{4.4}$$

The new coordinates are functions of the old ones, so by the product and chain rules for derivatives we find that

$$\frac{dx'^\mu}{d\tau} = \frac{\partial x'^\mu}{\partial x^\nu}\frac{dx^\nu}{d\tau} + \frac{\partial^2 x'^\mu}{\partial x^\sigma \partial x^\nu}\frac{dx^\sigma}{d\tau} x^\nu. \tag{4.5}$$

The first term is expected if the coordinate velocity is an order-1 tensor. But the second term, if nonzero, spoils the tensor transformation rule. That did not happen in the orthogonal-Galilean transformation, nor does it happen in any Lorentz transformation (why?). But it might happen in more general coordinate transformations.

What about the derivative of a vector component with respect to a coordinate, which arises in divergences and curls? It is left as an exercise to show that

$$(\partial_\lambda T^\alpha)' = \frac{\partial x'^\alpha}{\partial x^\rho}\frac{\partial x^\sigma}{\partial x'^\lambda}(\partial_\sigma T^\rho) + \frac{\partial^2 x'^\alpha}{\partial x^\rho \partial x^\sigma}\frac{\partial x^\sigma}{\partial x'^\lambda} T^\rho. \tag{4.6}$$

The first term on the right-hand side is that expected of a second-order tensor, but, again, the term with second derivatives may keep $\partial_\lambda T^\alpha$ from being a tensor of order 2.

In all the examples considered so far, second derivatives of new coordinates with respect to the old ones were encountered,

$$\frac{\partial^2 x'^{\alpha}}{\partial x^{\rho}\partial x^{\sigma}}, \tag{4.7}$$

which, if nonzero, means that the derivative of a tensor is not another tensor, preventing the derivative from being defined in coordinate-transcending, covariant language. This is serious for physics, whose fundamental equations are differential equations. It makes one worry whether "tensor calculus" might be a contradiction of terms!

An analogy may offer a useful geometric interpretation. In ordinary vector calculus the vector $\mathbf{v}(t)$ describing a particle's velocity is tangent to the trajectory swept out by the position vector $\mathbf{r}(t)$. In Euclidean geometries $\mathbf{v}$ and $\mathbf{r}$ live in the same space. But in a curved space this may not be so. Visualize the surface of a sphere, a two-dimensional curved space mapped with the coordinates of latitude and longitude. Two-dimensional inhabitants within the surface have no access to the third dimension. However, the velocity vector for a moving particle, as we outsiders are privileged to observe it from three-dimensional embedding space, is tangent to the spherical surface and points off it. But to the inhabitants of the surface, for something to exist it must lie *on* the surface. The notion of a tangent vector has no meaning for them other than locally, where a small patch of the sphere's surface cannot be distinguished from a Euclidean patch overlaid on it. If the globe's surface residents want to talk about velocity, they have to define it within the local Euclidean patch. Otherwise, for them the concept of velocity as a derivative does not exist. Similarly, in our world we may find that the derivative of a tensor is not always another tensor, suggesting that we could reside in some kind of curved space.

Since curvature requires a second derivative to be nonzero, these disturbing second derivatives that potentially spoil the transformation rules for tensors may be saying something about the curvature of a space. This problem never arose in Euclidean spaces or in the Minkowskian spacetime of special relativity, where the worrisome second derivatives all vanish. These spaces are said to be "flat" (see Ex. 4.1). But these questions become serious issues in the broader context of Riemannian and pseudo-Riemannian geometries, including important physics applications such as the general theory of relativity.

Let us demonstate the difficulty with a well-known application. It introduces a nontensor, called the "affine connection," that holds the key to resolving the problem.

4.2 The Affine Connection

Suppose that, as an adventuresome spirit, you take up skydiving. You leap out of the airplane, and, always the diligent physics student, you carry your trusty physics book with you. For the purposes of this discussion, suppose air resistance may be neglected before your parachute opens. While you are enjoying free fall, you let go of the physics book, and observe that it floats in front of you, as Galileo said it would. If you construct a set of coordinates centered on you, the location of the book relative to you could be described by a 4-vector with components X^μ. In your locally inertial frame, relative to you the book has zero acceleration,

$$\frac{d^2X^\alpha}{d\tau^2} = 0. \tag{4.8}$$

Let us transform this equation to the ground-based reference frame (assumed to be inertial) that uses coordinates x^μ relative to which the book accelerates. The skydiver frame and ground frame coordinates are related by some transformation

$$X^\mu = X^\mu(x^\nu). \tag{4.9}$$

Since each X^μ is a function of all the x^ν, Eq. (4.8) says

$$\frac{d}{d\tau}\left(\frac{\partial X^\alpha}{\partial x^\mu}\frac{dx^\mu}{d\tau}\right) = 0. \tag{4.10}$$

Using the chain and product rules for derivatives, this becomes

$$\frac{\partial X^\alpha}{\partial x^\mu}\frac{d^2x^\mu}{d\tau^2} + \frac{\partial^2 X^\alpha}{\partial x^\mu \partial x^\nu}\frac{dx^\mu}{d\tau}\frac{dx^\nu}{d\tau} = 0. \tag{4.11}$$

To reveal the book's acceleration relative to the ground frame by making $\frac{d^2x^\mu}{d\tau^2}$ stand alone, we need to get rid of the $\frac{\partial X^\alpha}{\partial x^\mu}$ coefficients, which we can readily do by making use of

$$\frac{\partial x^\lambda}{\partial X^\alpha}\frac{\partial X^\alpha}{\partial x^\mu} = \delta^\lambda{}_\mu. \tag{4.12}$$

To work this into our calculation, multiply Eq. (4.11) by $\frac{\partial x^\lambda}{\partial X^\alpha}$, so that

$$\frac{\partial x^\lambda}{\partial X^\alpha}\frac{\partial X^\alpha}{\partial x^\mu}\frac{d^2x^\mu}{d\tau^2} + \frac{\partial x^\lambda}{\partial X^\alpha}\frac{\partial^2 X^\alpha}{\partial x^\mu \partial x^\nu}\frac{dx^\mu}{d\tau}\frac{dx^\nu}{d\tau} = 0. \tag{4.13}$$

Recognizing the $\delta^\lambda{}_\mu$, we obtain

$$\frac{d^2x^\lambda}{d\tau^2} + \Gamma^\lambda{}_{\mu\nu}\frac{dx^\mu}{d\tau}\frac{dx^\nu}{d\tau} = 0, \tag{4.14}$$

where we have introduced the "affine connection,"

$$\Gamma^\lambda{}_{\mu\nu} \equiv \frac{\partial x^\lambda}{\partial X^\alpha}\frac{\partial^2 X^\alpha}{\partial x^\mu \partial x^\nu}, \tag{4.15}$$

which contains the worrisome second derivative of old coordinates with respect to the new ones. Under the typical circumstances (existence and continuity of the derivatives) where

$$\frac{\partial^2}{\partial x^\mu \partial x^\nu} = \frac{\partial^2}{\partial x^\nu \partial x^\mu}, \tag{4.16}$$

it follows that the affine connection, whatever it means, is symmetric in its subscripted indices,

$$\Gamma^{\lambda}{}_{\mu\nu} = \Gamma^{\lambda}{}_{\nu\mu}, \tag{4.17}$$

which we will always assume to be so (theories where this condition may not hold are said to have "torsion," but we won't need them).

Although the affine connection coefficients carry three indices, that does not guarantee them to be the components of a third-order tensor. The definitive test comes with its transformation under a change of coordinate system. This is an important question, and constructing the transformation for $\Gamma^{\lambda}{}_{\mu\nu}$ is simple in concept but can be tedious in practice.

Before going there, however, let us do some physics and see how this transformed free-fall equation compares to the Newtonian description of the same phenomenon. In so doing we might get a feel for the roles played in physics by the affine connection and the metric tensor, in particular, how they correspond, in some limit, to the Newtonian gravitational field.

4.3 The Newtonian Limit

We have seen that the equation describing a particle falling in a system of generalized coordinates, in which there is gravitation (and no other influences), is

$$\frac{d^2x^{\lambda}}{d\tau^2} + \Gamma^{\lambda}{}_{\mu\nu}u^{\mu}u^{\nu} = 0, \tag{4.18}$$

where

$$u^{\mu} \equiv \frac{dx^{\mu}}{d\tau} \tag{4.19}$$

denotes a component of the particle's 4-velocity in spacetime. This free-fall equation bears some resemblence to a component of the free-fall equation that comes from $\mathbf{F} = m\mathbf{a}$ when the only force is gravity, the familiar Newtonian $m\mathbf{g}$:

$$\frac{d^2\mathbf{a}}{dt^2} - \mathbf{g} = \mathbf{0}. \tag{4.20}$$

Comparing Eqs. (4.18) and (4.20), in both equations the first term is an acceleration, so evidently the Γ-term in Eq. (4.18) corresponds somehow to the gravitational field. Newtonian gravitation (as in solar system dynamics) is weak compared to that found near neutron stars or black holes (you can walk up a flight of stairs even though you are pulled downward by the entire Earth!). It is also essentially static since it deals with objects whose velocities are small compared to the speed of light. To connect Eq. (4.18) to Newtonian gravitation, we therefore take it to the weak-field, static limit. Besides offering a physical interpretation of the $\Gamma^{\lambda}{}_{\mu\nu}$ in the context of gravitation, this correspondence will also provide a "boundary condition" on a tensor theory of gravity.

In Minkowskian spacetime, the spacetime velocity components are

$$u^{\mu} = \gamma(1, \mathbf{v}), \tag{4.21}$$

where $\gamma = (1 - v^2)^{-1/2}$. If our falling particle moves with nonrelativistic speeds $v \ll 1$, then to first order in v, $\gamma \approx 1$ so that $u^0 = dt/d\tau \approx 1$, and $u^k \approx 0$ for $k = 1, 2, 3$. Under these circumstances Eq. (4.18) reduces to

$$\frac{d^2 x^{\lambda}}{dt^2} + \Gamma^{\lambda}{}_{00} \approx 0. \tag{4.22}$$

In Section 4.6 it will be shown that the $\Gamma^{\lambda}{}_{\mu\nu}$, in a given coordinate system, can be written in terms of that system's metric tensor, in particular,

$$\Gamma^{\lambda}{}_{\mu\nu} = \frac{1}{2} g^{\lambda\rho} \left[\frac{\partial g_{\rho\nu}}{\partial x^{\mu}} + \frac{\partial g_{\rho\mu}}{\partial x^{\nu}} - \frac{\partial g_{\mu\nu}}{\partial x^{\rho}} \right]. \tag{4.23}$$

In that case,

$$\Gamma^{\lambda}{}_{00} = \frac{1}{2} g^{\lambda\rho} \left[2 \frac{\partial g_{\rho 0}}{\partial x^0} - \frac{\partial g_{00}}{\partial x^{\rho}} \right]. \tag{4.24}$$

In a static or semi-static field,

$$\frac{\partial g_{\mu\nu}}{\partial x^0} \approx 0, \tag{4.25}$$

leaving $\Gamma^{\lambda}{}_{00} = -\frac{1}{2}\partial^{\lambda} g_{00}$. Furthermore, in a weak field we can write the metric tensor as the flat-spacetime Minkowski metric tensor $\eta_{\mu\nu}$ plus a small perturbation $h_{\mu\nu}$,

$$g_{\mu\nu} \approx \eta_{\mu\nu} + h_{\mu\nu}, \tag{4.26}$$

where the components of $\eta_{\mu\nu}$ are $0, +1$, or -1. Now to first order in the $h_{\mu\nu}$,

$$\Gamma^{\lambda}{}_{00} \approx -\frac{1}{2} \partial^{\lambda} h_{00}. \tag{4.27}$$

A static field eliminates the $\lambda = 0$ component, leaving from the spatial sector $\partial^k = -\partial/\partial x^k$. Thus, the surviving λ superscript denotes a spatial coordinate of index k, for which

$$\Gamma^{k}{}_{00} = \frac{1}{2} \frac{\partial h_{00}}{\partial x^k}. \tag{4.28}$$

Eq. (4.22) has become a component of

$$\frac{d^2 \mathbf{r}}{dt^2} = -\frac{1}{2} \boldsymbol{\nabla} h_{00}. \tag{4.29}$$

Compare this to $\mathbf{F} = m\mathbf{a}$ with $\mathbf{F}$ identified as $m\mathbf{g}$. In terms of the Newtonian potential ϕ, where $\mathbf{g} = -\boldsymbol{\nabla}\phi$, we see that

$$h_{00} = 2\phi. \tag{4.30}$$

Since $g_{00} = \eta_{00} + h_{00} = 1 + h_{00}$, the Newtonian-limit connection between one component of the metric tensor and the nonrelativistic equations of free fall has been identified:

$$(g_{00})_{Nt\ limit} \approx 1 + 2\phi. \tag{4.31}$$

For a point mass m, in conventional units this becomes

$$(g_{00})_{Nt\ limit} \approx 1 - \frac{2Gm}{rc^2}. \tag{4.32}$$

4.4 Transformation of the Affine Connection

Is $\Gamma^\lambda{}_{\mu\nu}$ a tensor? To approach this question, consider *two* ground frames and compare their affine connection coefficients to one other, each obtained by tracking the skydiver book acceleration. How does the affine connection that relates the X^μ to the x^μ compare to the affine connection that connects the X^μ to some other coordinate system x'^μ? Let the ground frame 1 origin be located at the radar screen of the airport control tower, and let the ground frame 2 origin be located at the base of the town square's clock tower. We must transform Eq. (4.15) from the unprimed coordinates (control tower origin) to the primed frame coordinates (town square origin) and see whether $\Gamma^\lambda{}_{\mu\nu}$ respects the tensor transformation rule for a tensor of order 3.

In writing the affine connection in the town square frame, the x^μ get replaced with x'^μ while leaving the skydiver frame coordinates X^α alone. Denoting $\Gamma^{\lambda'}{}_{\mu'\nu'} \equiv \Gamma'^\lambda{}_{\mu\nu}$, we now examine the transformation properties of

$$\Gamma'^\lambda{}_{\mu\nu} = \left(\frac{\partial x'^\lambda}{\partial X^\alpha}\right)\left[\frac{\partial^2 X^\alpha}{\partial x'^\mu \partial x'^\nu}\right]. \tag{4.33}$$

Since the x'^λ and the X^α are functions of the x^μ, two applications of the chain rule give

$$\Gamma'^\lambda{}_{\mu\nu} = \left(\frac{\partial x'^\lambda}{\partial x^\rho}\frac{\partial x^\rho}{\partial X^\alpha}\right)\frac{\partial}{\partial x'^\mu}\left[\frac{\partial X^\alpha}{\partial x^\sigma}\frac{\partial x^\sigma}{\partial x'^\nu}\right]. \tag{4.34}$$

Pull the derivative with respect to x'^μ through the square-bracketed terms, to obtain

$$\Gamma'^\lambda{}_{\mu\nu} = \left(\frac{\partial x'^\lambda}{\partial x^\rho}\frac{\partial x^\rho}{\partial X^\alpha}\right)\left[\frac{\partial^2 X^\alpha}{\partial x^\tau \partial x^\sigma}\frac{\partial x^\tau}{\partial x'^\mu}\frac{\partial x^\sigma}{\partial x'^\nu} + \frac{\partial X^\alpha}{\partial x^\sigma}\frac{\partial^2 x^\sigma}{\partial x'^\mu \partial x'^\nu}\right]. \tag{4.35}$$

Examine closely the first term on the right-hand side of this equation,

$$\left(\frac{\partial x'^\lambda}{\partial x^\rho}\frac{\partial x^\rho}{\partial X^\alpha}\right)\left[\frac{\partial^2 X^\alpha}{\partial x^\tau \partial x^\sigma}\frac{\partial x^\tau}{\partial x'^\mu}\frac{\partial x^\sigma}{\partial x'^\nu}\right], \tag{4.36}$$

which can be regrouped as

$$\left(\frac{\partial x'^{\lambda}}{\partial x^{\rho}}\frac{\partial x^{\tau}}{\partial x'^{\mu}}\frac{\partial x^{\sigma}}{\partial x'^{\nu}}\right)\left(\frac{\partial x^{\rho}}{\partial X^{\alpha}}\frac{\partial^2 X^{\alpha}}{\partial x^{\tau}\partial x^{\sigma}}\right) \tag{4.37}$$

and recognized to be

$$\frac{\partial x'^{\lambda}}{\partial x^{\rho}}\frac{\partial x^{\tau}}{\partial x'^{\mu}}\frac{\partial x^{\sigma}}{\partial x'^{\nu}}\Gamma^{\rho}{}_{\tau\sigma}. \tag{4.38}$$

The second term in Eq. (4.35) contains

$$\left(\frac{\partial x'^{\lambda}}{\partial x^{\rho}}\frac{\partial x^{\rho}}{\partial X^{\alpha}}\right)\left[\frac{\partial X^{\alpha}}{\partial x^{\sigma}}\frac{\partial^2 x^{\sigma}}{\partial x'^{\mu}\partial x'^{\nu}}\right] \tag{4.39}$$

and can be regrouped as

$$\frac{\partial x'^{\lambda}}{\partial x^{\rho}}\left(\frac{\partial x^{\rho}}{\partial X^{\alpha}}\frac{\partial X^{\alpha}}{\partial x^{\sigma}}\right)\frac{\partial^2 x^{\sigma}}{\partial x'^{\mu}\partial x'^{\nu}}. \tag{4.40}$$

The part between parentheses contains the Kronecker delta,

$$\frac{\partial x^{\rho}}{\partial X^{\alpha}}\frac{\partial X^{\alpha}}{\partial x^{\sigma}} = \delta^{\rho}{}_{\sigma}, \tag{4.41}$$

which leaves

$$\frac{\partial x'^{\lambda}}{\partial x^{\rho}}\frac{\partial^2 x^{\rho}}{\partial x'^{\mu}\partial x'^{\nu}}. \tag{4.42}$$

Now when everything gets put back together, we find that the affine connection transforms according to

$$\Gamma'^{\lambda}{}_{\mu\nu} = \frac{\partial x'^{\lambda}}{\partial x^{\rho}}\frac{\partial x^{\tau}}{\partial x'^{\mu}}\frac{\partial x^{\sigma}}{\partial x'^{\nu}}\Gamma^{\rho}{}_{\tau\sigma} + \frac{\partial x'^{\lambda}}{\partial x^{\rho}}\frac{\partial^2 x^{\rho}}{\partial x'^{\mu}\partial x'^{\nu}}. \tag{4.43}$$

The first term after the equals sign would be the end of the story if the affine connection coefficients were the components of a third-order tensor with one upstairs and two downstairs indices. However, the additional term (which looks like another affine connection relating the control tower and town square frames), if not zero, prevents the affine connection from transforming as a tensor.

Fortunately, the term that spoils the affine connection's third-order tensor candidacy contains a second derivative similar to that which prevented the derivative of a tensor from being another tensor (recall Section 4.1). This observation offers hope for salvaging the situation. Perhaps we can extend what we mean by "derivative" and add to the usual derivative a Γ-term, such that, when carrying out the transformation, the offending terms contributed by the usual derivative and by the Γ-term cancel out. We will try it in the next section.

In the transformation of the affine connection, the last term in Eq. (4.43), the one that prevents $\Gamma^{\lambda}{}_{\mu\nu}$ from being a tensor, is a second derivative of an *old* coordinate with respect to *new* ones, $\partial^2 x^\rho/\partial x'^\mu \partial x'^\nu$. This term can alternatively be written as the second derivative of the *new* coordinates with respect to the *old* ones, $\partial^2 x'^\rho/\partial x^\mu \partial x^\nu$. To demonstrate this, begin with the identity

$$\frac{\partial x'^\lambda}{\partial x^\rho}\frac{\partial x^\rho}{\partial x'^\nu} = \delta^\lambda{}_\nu \tag{4.44}$$

and differentiate it with respect to x'^μ:

$$\left(\frac{\partial^2 x'^\lambda}{\partial x^\rho \partial x^\sigma}\frac{\partial x^\sigma}{\partial x'^\mu}\right)\frac{\partial x^\rho}{\partial x'^\nu} + \frac{\partial x'^\lambda}{\partial x^\rho}\left(\frac{\partial^2 x^\rho}{\partial x'^\mu \partial x'^\nu}\right) = 0. \tag{4.45}$$

Recognizing the second term as also appearing in Eq. (4.43), we can rewrite the transformation of the affine connection an alternate way,

$$\Gamma'^\lambda{}_{\mu\nu} = \frac{\partial x'^\lambda}{\partial x^\rho}\frac{\partial x^\tau}{\partial x'^\mu}\frac{\partial x^\sigma}{\partial x'^\nu}\Gamma^\rho{}_{\tau\sigma} - \frac{\partial x^\sigma}{\partial x'^\mu}\frac{\partial x^\rho}{\partial x'^\nu}\frac{\partial^2 x'^\lambda}{\partial x^\rho \partial x^\sigma}. \tag{4.46}$$

Incidentally, even though the affine connection coefficients are not the components of tensors, their indices can be raised and lowered in the usual way, such as

$$\begin{aligned} g_{\rho\lambda}\Gamma^\lambda{}_{\mu\nu} &= \Gamma_{\rho\mu\nu} \\ &= \frac{1}{2}g_\rho{}^\sigma[\partial_\mu g_{\nu\sigma} + \partial_\nu g_{\mu\sigma} - \partial_\sigma g_{\mu\nu}]. \end{aligned}$$

That is why the superscript does not stand above the subscripts in $\Gamma^\lambda{}_{\mu\nu}$.

4.5 The Covariant Derivative

In Section 4.1 we saw that the derivative of a vector is not necessarily another vector. Let us isolate the anomalous term that prevents $dx^\mu/d\tau$ from transforming as a vector:

$$\frac{dx'^\mu}{d\tau} - \frac{\partial x'^\mu}{\partial x^\nu}\frac{dx^\nu}{d\tau} = \left(\frac{\partial^2 x'^\mu}{\partial x^\rho \partial x^\nu}\right)\frac{dx^\rho}{d\tau}\frac{dx^\nu}{d\tau}. \tag{4.47}$$

Similarly, the anomaly that prevents the derivative of a vector component with respect to a coordinate from transforming as a second-order tensor can be isolated:

$$T'^\mu{}_{,\nu} - \frac{\partial x'^\mu}{\partial x^\rho}\frac{\partial x^\sigma}{\partial x'^\nu}T^\rho{}_{,\sigma} = \left(\frac{\partial^2 x'^\mu}{\partial x^\rho \partial x^\sigma}\right)\frac{\partial x^\sigma}{\partial x'^\nu}T^\rho, \tag{4.48}$$

where we recall the notation $\partial_\nu T^\mu \equiv T^\mu{}_{,\nu}$ (a "comma as derivative" *precedes* the index, so that "$A_{,\mu}$" means $\partial A/\partial x^\mu$ whereas "A_μ," means an ordinary punctuation comma). From Section 4.4 we can also isolate the factor that prevents the affine connection from being a third-order tensor:

$$\Gamma'^{\lambda}{}_{\mu\nu} - \frac{\partial x'^{\lambda}}{\partial x^{\rho}} \frac{\partial x^{\tau}}{\partial x'^{\mu}} \frac{\partial x^{\sigma}}{\partial x'^{\nu}} \Gamma^{\rho}{}_{\tau\sigma} = -\left(\frac{\partial^2 x'^{\lambda}}{\partial x^{\rho} \partial x'^{\sigma}}\right) \frac{\partial x^{\sigma}}{\partial x'^{\mu}} \frac{\partial x^{\rho}}{\partial x'^{\nu}}. \tag{4.49}$$

In each of these cases, the offending term includes a second derivative of the new coordinates with respect to the old ones. If the definition of what we mean by "derivative" was extended by adding to the usual derivative a term that includes the affine connection, then hopefully the nontensor parts of both might cancel out! By experimenting with the factors to be added, one finds that the following "re-defined derivatives" transform as tensors. Called "covariant derivatives," they are dignified with several notations such as D or a semicolon, which will be introduced along with examples.

The covariant derivative of a vector component with respect to a scalar is

$$\frac{dA^{\lambda}}{d\tau} + \Gamma^{\lambda}{}_{\mu\nu} \frac{dx^{\mu}}{d\tau} A^{\nu} \equiv \frac{DA^{\lambda}}{D\tau}. \tag{4.50}$$

The covariant derivative of a vector component with respect to a coordinate is

$$A^{\lambda}{}_{,\mu} + \Gamma^{\lambda}{}_{\mu\nu} A^{\nu} \equiv D_{\mu} A^{\lambda} \equiv A^{\lambda}{}_{;\mu}. \tag{4.51}$$

The covariant derivative of a dual vector component is

$$A_{\lambda,\mu} - \Gamma^{\alpha}{}_{\lambda\mu} A_{\alpha} \equiv D_{\mu} A_{\lambda} \equiv A_{\lambda;\mu}. \tag{4.52}$$

This notation for expanding "derivative" into "covariant derivative" suggests an algorithm: "Replace d or ∂ with D" or, as stated in another notation, "Replace the comma with a semicolon." Conceptually, if not precisely, $D \equiv \partial + \Gamma - term$. In the comma-to-semicolon notation, $T_{;\mu} \equiv T_{,\mu} + \Gamma - term$. (Some authors use $\boldsymbol{\nabla}$ to denote the covariant derivative, but in this book $\boldsymbol{\nabla}$ is the familiar Euclidean gradient.)

Now if we return to our freely falling skydiver, in the skydiver frame the "no acceleration" condition may be stated

$$\frac{dU^{\lambda}}{d\tau} = 0, \tag{4.53}$$

where $U^{\lambda} = dX^{\lambda}/d\tau$. To switch to the ground frame, replace the local X^{μ} coordinates with the global x^{μ} coordinates, replace the local metric tensor $\eta_{\mu\nu}$ with the global $g_{\mu\nu}$, and replace the "d" derivative with the covariant "D" derivative. Now the "no acceleration" condition gets rewritten as

$$\frac{Du^{\lambda}}{D\tau} = 0. \tag{4.54}$$

Writing out what the covariant derivative means, we recognize the result (compare to Eq. (4.18):

$$\frac{du^\lambda}{d\tau} + \Gamma^\lambda{}_{\mu\nu}\frac{dx^\mu}{d\tau}u^\nu = 0. \tag{4.55}$$

The covariant derivatives are allegedly *defined* such that they transform as tensors, and that claim must be demonstrated before we can put confidence in these definitions. Consider the covariant derivative of a vector with respect to a scalar, which in the system of primed coordinates reads

$$\left(\frac{DA^\lambda}{D\tau}\right)' = \frac{dA'^\lambda}{d\tau} + \Gamma'^\lambda{}_{\mu\nu}\frac{dx'^\mu}{d\tau}A'^\nu. \tag{4.56}$$

When the primed quantities on the right hand-side of this equation are written in terms of transformations from the unprimed ones, and some Kronecker deltas are recognized, one shows (Ex. 4.10) that

$$\left(\frac{DA^\lambda}{d\tau}\right)' = \frac{\partial x'^\lambda}{\partial x^\rho}\frac{DA^\rho}{D\tau}, \tag{4.57}$$

which is the tensor transformation rule for a vector. Similarly, one can show in a straightforward if tedious manner (Ex. 4.11) that the covariant derivative of a vector with respect to a coordinate is a second-order tensor, because

$$(A^\lambda{}_{;\mu})' = \frac{\partial x'^\lambda}{\partial x^\rho}\frac{\partial x^\sigma}{\partial x'^\mu}A^\rho{}_{;\sigma}. \tag{4.58}$$

A few covariant derivatives of higher-order tensors are listed below:

$$T^{\mu\nu}{}_{;\beta} = T^{\mu\nu}{}_{,\beta} + \Gamma^\mu{}_{\alpha\beta}T^{\alpha\nu} + \Gamma^\nu{}_{\alpha\beta}T^{\mu\alpha} \tag{4.59}$$

$$S^{\mu\nu}{}_{\sigma;\beta} = S^{\mu\nu}{}_{\sigma,\beta} + \Gamma^\mu{}_{\alpha\beta}S^{\alpha\nu}{}_\sigma + \Gamma^\nu{}_{\alpha\beta}S^{\mu\alpha}{}_\sigma - \Gamma^\alpha{}_{\sigma\beta}S^{\mu\nu}{}_\alpha. \tag{4.60}$$

One sees a pattern in the addition of $+\Gamma$ terms for derivatives of tensors with upstairs indices, and $-\Gamma$ terms for the derivatives of tensors bearing downstairs indices.

4.6 Relation of the Affine Connection to the Metric Tensor

In a transformation of coordinates, the affine connection coefficients bridge the two systems according to

$$\Gamma^\lambda{}_{\mu\nu} = \frac{\partial x^\lambda}{\partial x'^\rho}\frac{\partial^2 x'^\rho}{\partial x^\mu \partial x^\nu}. \tag{4.61}$$

But when solving a problem, one typically works *within* a given coordinate system. It would be advantageous if the $\Gamma^{\lambda}{}_{\mu\nu}$ could be computed from within a single coordinate system. This can be done. Within a coordinate system that has metric tensor $g_{\mu\nu}$, the affine connection coefficients will be shown to be

$$\Gamma^{\lambda}{}_{\mu\nu} = \frac{1}{2} g^{\lambda\rho}\left[\partial_\mu g_{\nu\rho} + \partial_\nu g_{\mu\rho} - \partial_\rho g_{\mu\nu}\right]. \tag{4.62}$$

Like most of the important points in this subject, the proof that the affine connection coefficients can be calculated this way is messy to demonstrate. Since $\Gamma^{\lambda}{}_{\mu\nu}$ bridges the old and new coordinate systems, start with the metric tensor transformed into new coordinates,

$$g'_{\mu\nu} = \frac{\partial x^\rho}{\partial x'_\mu}\frac{\partial x^\sigma}{\partial x'^\nu}\, g_{\rho\sigma}, \tag{4.63}$$

and evaluate its (non-covariant) derivative with respect to a new coordinate:

$$\frac{\partial g'_{\mu\nu}}{\partial x'^\kappa} = \frac{\partial}{\partial x'^\kappa}\left(\frac{\partial x^\rho}{\partial x'^\mu}\frac{\partial x^\sigma}{\partial x'^\nu} g_{\rho\sigma}\right). \tag{4.64}$$

Now pull the derivative through, using the product and chain rules for derivatives. We confront a huge mess:

$$\partial_{\kappa'} g'_{\mu\nu} = \left(\frac{\partial^2 x^\rho}{\partial x'^\kappa \partial x'^\mu}\right)(\partial_{\nu'} x^\sigma) g_{\rho\sigma} + (\partial_{\mu'} x^\rho)\left(\frac{\partial^2 x'^\sigma}{\partial x'^\kappa \partial x'^\nu}\right) g_{\rho\sigma} + (\partial_{\mu'} x^\rho)(\partial_{\nu'} x^\sigma)(\partial_\tau g_{\rho\sigma})(\partial_{\kappa'} x^\tau). \tag{4.65}$$

Let us temporarily drop the primes and consider the permutations of the indices on $\partial_\kappa g_{\mu\nu}$. This presents us with two more terms, $\partial_\mu g_{\nu\kappa}$ and $\partial_\nu g_{\kappa\mu}$. From these permutations construct the quantity

$$\frac{1}{2}\left[\partial_\mu g_{\kappa\nu} + \partial_\nu g_{\kappa\mu} - \partial_\kappa g_{\mu\nu}\right] \equiv [\mu\nu, \kappa], \tag{4.66}$$

which is called the "Christoffel symbol of the first kind." Multiply it by a component of the metric tensor with upper indices and do a contraction, to define the "Christoffel symbol of the second kind,"

$$\begin{aligned} \left\{ \begin{matrix} & \lambda & \\ \mu & & \nu \end{matrix} \right\} &\equiv g^{\lambda\kappa}[\mu\nu, \kappa] \\ &= \frac{1}{2} g^{\lambda\kappa}\left[\partial_\mu g_{\kappa\nu} + \partial_\nu g_{\kappa\mu} - \partial_\kappa g_{\mu\nu}\right]. \end{aligned}$$

Now go back to Eq. (4.65), restore the primes, and, as suggested, rewrite it two more times with permuted indices. Add two of the permutations and subtract the third one, and then contract with $g^{\lambda\kappa}$, which leads to

$$\left\{ \begin{matrix} & \lambda & \\ \mu & & \nu \end{matrix} \right\}' = \frac{\partial x'^\lambda}{\partial x^\kappa}\frac{\partial x^\tau}{\partial x'^\mu}\frac{\partial x^\sigma}{\partial x'^\nu}\left\{ \begin{matrix} & \kappa & \\ \tau & & \sigma \end{matrix} \right\} + \frac{\partial^2 x^\kappa}{\partial x'^\mu \partial x'^\nu}\frac{\partial x'^\lambda}{\partial x^\kappa}. \tag{4.67}$$

This is precisely the same transformation we saw for $\Gamma^\lambda{}_{\mu\nu}$, Eq. (4.43)! Thus, the *difference* between the affine connection and the Christoffel symbol respects the transformation rule for third-order tensors, because the troublesome second derivative cancels:

$$\left[\Gamma^\lambda{}_{\mu\nu} - \left\{ \begin{matrix} \lambda \\ \mu \quad \nu \end{matrix} \right\}\right]' = \frac{\partial x'^\lambda}{\partial x^\rho}\frac{\partial x^\tau}{\partial x'^\mu}\frac{\partial x^\sigma}{\partial x'^\nu}\left[\Gamma^\rho{}_{\tau\sigma} - \left\{ \begin{matrix} \rho \\ \tau \quad \sigma \end{matrix} \right\}\right]. \tag{4.68}$$

Therefore, whatever the difference between the affine connection and Christoffel coefficients happens to be, it will be the same in all frames that are connected by a family of allowed transformations. In particular, in general relativity, we saw that the affine connection coefficients correspond to the presence of a gravitational field, which vanishes in the local free-fall frame (see Ex. 4.1). Also, in that free-fall frame, the metric tensor components are those of "flat" spacetime, where all the $\left\{ \begin{matrix} \lambda \\ \mu \quad \nu \end{matrix} \right\}$ vanish there as well. Thus, $\Gamma^\lambda{}_{\mu\nu} - \left\{ \begin{matrix} \lambda \\ \mu \quad \nu \end{matrix} \right\} = 0$ in the free-fall frame, and by general covariance $\Gamma^\lambda{}_{\mu\nu} - \left\{ \begin{matrix} \lambda \\ \mu \quad \nu \end{matrix} \right\} = 0$ in all reference frames whose coordinates can be derived from those of the free-fall system by a transformation. Thus, we can say, at least in the spacetimes of general relativity, that

$$\Gamma^\lambda{}_{\mu\nu} = \frac{1}{2}g^{\lambda\kappa}[\partial_\mu g_{\nu\kappa} + \partial_\nu g_{\mu\kappa} - \partial_\kappa g_{\mu\nu}]. \tag{4.69}$$

One admires the patience and persistence of the people who originally worked all of this out. See Section 7.4 for an alternative introduction to the $\Gamma^\lambda{}_{\mu\nu}$ that relies on the properties of basis vectors.

4.7 Divergence, Curl, and Laplacian with Covariant Derivatives

Here we resume the discussion of Section 3.5 concerning the relation between "ordinary" vector components on the one hand and contravariant and covariant vector components on the other hand. In this section we apply this lens to derivatives built from the gradient, including the divergence, the curl, and the Laplacian.

Let $\hat{\mathbf{e}}_\mu$ denote a unit basis vector, in Euclidean space, pointing in the direction of increasing coordinate x^μ. An "ordinary vector" takes the following generic form in rectangular, spherical, and cylindrical coordinates in Euclidean space:

$$\tilde{\mathbf{A}} = \tilde{A}_1\hat{\mathbf{e}}_1 + \tilde{A}_2\hat{\mathbf{e}}_2 + \tilde{A}_3\hat{\mathbf{e}}_3, \tag{4.70}$$

where the subscripts on the coefficients $\tilde{A}_\mu$ and $\hat{\mathbf{e}}_\mu$ have no covariant/contravariant meaning; they are merely labels. The point of Section 3.5 was to discuss the relationship of the $\tilde{A}_\mu$ with contravariant and covariant first-order tensor components. In these Euclidean coordinate systems, the metric tensor took the diagonal form $g_{\mu\nu} = (h_\mu)^2\delta_{\mu\nu}$ (no sum). We determined that, for contravariant first-order tensors (a.k.a. "vectors"),

$$A^\mu = \frac{\tilde{A}_\mu}{h_\mu}, \tag{4.71}$$

and for the covariant order-1 tensors (a.k.a. "dual vectors"),

$$A_\mu = h_\mu \tilde{A}_\mu \tag{4.72}$$

(no sum). In addition, and consistent with this, we saw that the components of the "ordinary" gradient of a scalar function, $\boldsymbol{\nabla} f$, are

$$\nabla_\mu f = \frac{1}{h_\mu}\partial_\mu f \tag{4.73}$$

(no sum), which makes explicit the distinction between the derivative with respect to a *coordinate* (to give $\partial_\mu f$, a first-order tensor) and the derivative with respect to a *distance* (that is, $\nabla_\mu f$).

However, in Riemannian and pseudo-Riemannian spaces, to be a tensor the extension of the "ordinary" gradient $\boldsymbol{\nabla}$ must be not only written in terms of the ∂_μ but subsequently generalized to covariant derivatives D_μ as well. In this way the "ordinary" divergence $\boldsymbol{\nabla} \cdot \tilde{\mathbf{A}}$ gets replaced with

$$D_\mu A^\mu \equiv A^\mu{}_{;\mu} = \partial_\mu A^\mu + \Gamma^\mu{}_{\mu\lambda} A^\lambda \tag{4.74}$$

(restoring the summation convention). The contracted affine connection is, by Eq. (4.62),

$$\Gamma^\mu{}_{\mu\lambda} = \frac{1}{2} g^{\mu\rho}(\partial_\lambda g_{\rho\mu}). \tag{4.75}$$

In addition, with the result of Ex. 4.15, where one shows that

$$\Gamma^\mu{}_{\mu\lambda} = \frac{1}{\sqrt{|g|}}\partial_\lambda \sqrt{|g|}, \tag{4.76}$$

the covariant divergence may be neatly written

$$D_\mu A^\mu = \partial_\mu A^\mu + \frac{1}{\sqrt{|g|}}\left(\partial_\lambda \sqrt{|g|}\right) A^\lambda. \tag{4.77}$$

As one more simplification, by the product rule for derivatives one notices that

$$\partial_\lambda(\sqrt{|g|}A^\lambda) = (\partial_\lambda \sqrt{|g|})A^\lambda + \sqrt{|g|}(\partial_\lambda A^\lambda), \tag{4.78}$$

whereby the covariant divergence becomes the tidy expression

$$A^{\mu}{}_{;\mu} = \frac{1}{\sqrt{|g|}}\partial_\mu(\sqrt{|g|}A^\mu). \tag{4.79}$$

As a consistency check, the covariant divergence should reduce to the "ordinary" divergence when the Riemannian space is good old Euclidean geometry mapped in spherical coordinates $(x^1, x^2, x^3) = (r, \theta, \varphi)$, for which $\sqrt{|g|} = r^2 \sin\theta$. Eq. (4.79) indeed yields the familiar result for the "ordinary" divergence,

$$\begin{aligned}(A^{\mu}{}_{;\mu})_{Euclidean\ (r,\theta,\varphi)} &= \frac{1}{r^2\sin\theta}\left[\frac{\partial(r^2\sin\theta A^1)}{\partial x^1} + \frac{\partial(r^2\sin\theta A^2)}{\partial x^2} + \frac{\partial(r^2\sin\theta A^3)}{\partial x^3}\right] \\ &= \frac{1}{r^2}\frac{\partial(r^2\tilde{A}_r)}{\partial r} + \frac{1}{r\sin\theta}\frac{\partial}{\partial\theta}(\sin\theta\tilde{A}_\theta) + \frac{1}{r\sin\theta}\frac{\partial A_\varphi}{\partial\varphi} \\ &= \boldsymbol{\nabla}\cdot\tilde{\mathbf{A}},\end{aligned}$$

where we have recalled $A^\mu = \tilde{A}_\mu/h_\mu$ (no sum).

Now let us extend the "ordinary curl" to the "covariant derivative curl." The "ordinary curl" is typically defined in rectangular coordinates, for which the ith component of $\tilde{\mathbf{C}} = \boldsymbol{\nabla}\times\tilde{\mathbf{A}}$ reads

$$\tilde{C}_i = \partial_j\tilde{A}_k - \partial_k\tilde{A}_j, \tag{4.80}$$

where ijk appear in cyclic order. In spherical coordinates,

$$\tilde{\mathbf{C}} = \boldsymbol{\nabla}\times\tilde{\mathbf{A}} = \tilde{C}_r\hat{\mathbf{r}} + \tilde{C}_\theta\hat{\theta} + \tilde{C}_\varphi\hat{\varphi}, \tag{4.81}$$

where

$$\tilde{C}_r = \frac{1}{\sin\theta}\partial_\theta(\sin\theta\tilde{A}_\varphi) \tag{4.82}$$

$$\tilde{C}_\theta = \frac{1}{r}\left[\frac{1}{\sin\theta}\partial_\varphi\tilde{A}_r - \partial_r(r\tilde{A}_\varphi)\right] \tag{4.83}$$

$$\tilde{C}_\varphi = \frac{1}{r}\left[\partial_r(r\tilde{A}_\theta) - \partial_\theta\tilde{A}_r\right]. \tag{4.84}$$

A candidate extension of the curl, in the coordinate systems based on Eq. (4.80) (still Euclidean space), would be to replace the ordinary curl component $\tilde{C}_i$ with C_μ, to define a curl in generalized coordinates according to

$$C_\alpha = \partial_\beta A_\gamma - \partial_\gamma A_\beta, \tag{4.85}$$

where $\alpha\beta\gamma$ are integer labels in cyclic order, and $A_\mu = h_\mu\tilde{A}_\mu$ (no sum). Noting that, in spherical coordinates, $\sqrt{|g|} = r^2\sin\theta$, $h_1 = 1$, $h_2 = r$, and $h_3 = r\sin\theta$, in the instance of $\mu = 1$ this would be

$$\begin{aligned} C_1 &= \partial_2 A_3 - \partial_3 A_2 \\ &= \partial_\theta (r \sin\theta \tilde{A}_\varphi) - \partial_\varphi (r \tilde{A}_\theta) \\ &= r^2 \sin\theta \tilde{C}_r \\ &= \sqrt{|g|}\frac{\tilde{C}_r}{1}. \end{aligned}$$

Similarly, after considering $\mu = 2$ and $\mu = 3$, we establish the relation between components of the "ordinary" curl and the curl made from covariant vector components,

$$C_\mu = \sqrt{|g|}\frac{\tilde{C}_\mu}{h_\mu} \tag{4.86}$$

(no sum).

Now we can go beyond Euclidean spaces and inquire about a "covariant derivative curl" by adding to the partial derivatives the corresponding affine connection terms. Extend Eq. (4.85) to a candidate covariant derivative curl

$$C_\alpha = D_\beta A_\gamma - D_\gamma A_\beta. \tag{4.87}$$

Because $\Gamma^\lambda{}_{\mu\nu} = \Gamma^\lambda{}_{\nu\mu}$, the affine connection terms cancel out,

$$D_\beta A_\gamma - D_\gamma A_\beta = \partial_\beta A_\gamma - \partial_\gamma A_\beta, \tag{4.88}$$

leaving Eq. (4.85) as the correct curl with covariant derivatives. However, this looks strange, since C_α on the left-hand side of Eq. (4.85) appears notationally like a tensor of order 1, whereas $\partial_\beta A_\gamma - \partial_\gamma A_\beta$ on the right-hand side is a second-order tensor. Sometimes it is convenient to use one index and a supplementary note about cyclic order; in other situations it will be more convenient to rename the left-hand side of Eq. (4.85) according to

$$C_{\beta\gamma} \equiv \partial_\beta A_\gamma - \partial_\gamma A_\beta, \tag{4.89}$$

which will serve as our tensor definition of the curl, an order-2 tensor. Notice that the covariant curl, like the "ordinary" curl, is antisymmetric, $C_{\mu\nu} = -C_{\nu\mu}$.

By similar reasoning, the Laplacian made of covariant derivatives can be defined by starting with the definition of the "ordinary" one as the divergence of a gradient,

$$\nabla^2\phi = \boldsymbol{\nabla}\cdot\boldsymbol{\nabla}\phi, \tag{4.90}$$

which inspires the tensor version

$$\partial_\mu \partial^\mu \phi, \tag{4.91}$$

in which the usual derivatives are next replaced with covariant derivatives, $D_\mu D^\mu \phi$. From this it follows that

$$(D^\mu \phi)_{;\mu} = \frac{1}{\sqrt{|g|}}\partial_\mu\left[\sqrt{|g|}\partial^\mu\phi\right], \tag{4.92}$$

which is consistent with Eq. (4.79).

4.8 Discussion Questions and Exercises

Discussion Questions

Q4.1 In thermodynamics, the change in internal energy ΔU is related to the heat Q put into the system and the work W done by the system, according to $\Delta U = Q - W$, the first law of thermodynamics. For any thermodynamic process, Q and W are path dependent in the space of thermodynamic variables, but ΔU is path independent. Describe how this situation offers an analogy to tensor calculus, where a derivative plus an affine connection term yields a tensor, even though the derivative and the affine connection are not separately tensors.

Q4.2 In the context of general relativity, the affine connection coefficients correspond to the presence of a gravitational field. How can this correspondence be justified?

Exercises

4.1 Show that

$$\frac{\partial^2 x'^\lambda}{\partial x^\mu \partial x^\nu} = 0 \tag{4.93}$$

for a Lorentz transformation between Minkowskian coordinate systems.

4.2 (a) Show that the covariant derivative of the metric tensor vanishes, $g_{\mu\nu;\alpha} = 0$. Hint: consider interchanging summed dummy indices.
(b) Show also that $g^{\mu\nu}{}_{;\alpha} = 0$.

4.3 In spacetime mapped with coordinates $x^0 = t$, $x^1 = r$, $x^2 = \theta$, and $x^3 = \varphi$, consider a metric tensor of the form (see Ohanian, p. 229)

$$g_{\mu\nu} = \begin{pmatrix} e^{A(r)} & 0 & 0 & 0 \\ 0 & -e^{B(r)} & 0 & 0 \\ 0 & 0 & -r^2 & 0 \\ 0 & 0 & 0 & r^2\sin^2\theta \end{pmatrix}. \tag{4.94}$$

(a) Show that the nonzero affine connection coefficients are

$$\Gamma^0{}_{01} = \frac{1}{2}A' \tag{4.95}$$

$$\Gamma^1{}_{00} = \frac{1}{2}A' e^{A-B} \tag{4.96}$$

$$\Gamma^1{}_{11} = \frac{1}{2}B' \tag{4.97}$$

$$\Gamma^1{}_{22} = -re^{-B} \tag{4.98}$$

$$\Gamma^1{}_{33} = -re^{-B}\sin^2\theta \tag{4.99}$$

$$\Gamma^2{}_{12} = \Gamma^3{}_{13} = \frac{1}{r} \tag{4.100}$$

$$\Gamma^2{}_{33} = -\sin\theta\cos\theta \tag{4.101}$$

$$\Gamma^3{}_{23} = -\cot\theta, \tag{4.102}$$

where the primes denote the derivative with respect to r (for a shorter version of this problem, do the first three coefficients).
(b) What do these affine connection coefficients become in Minkowskian spacetime?

4.4 In the same coordinate system as Ex. 4.3, consider the time-dependent, off-diagonal metric tensor with parameters $A(t,r), B(t,r)$, and $C(t,r)$,

$$g_{\mu\nu} = \begin{pmatrix} A(t,r) & C(t,r) & 0 & 0 \\ C(t,r) & -B(t,r) & 0 & 0 \\ 0 & 0 & -r^2 & 0 \\ 0 & 0 & 0 & r^2\sin^2\theta \end{pmatrix}.$$

(a) Construct $g^{\mu\nu}$.
(b) Let primes denote the partial derivative with respect to r, and overdots the partial derivative with respect to t. With $D \equiv AB + C^2$, show that the nonzero affine connection coefficients are

$$\Gamma^0{}_{00} = \frac{1}{2D}(\dot{A}B - CA' + 2C\dot{C}) \tag{4.103}$$

$$\Gamma^0{}_{01} = \frac{1}{2D}(A'B - C\dot{B}) \tag{4.104}$$

$$\Gamma^0{}_{11} = \frac{1}{2D}(2BC' - CB' + B\dot{B}) \tag{4.105}$$

$$\Gamma^0{}_{22} = \frac{rC}{D} \tag{4.106}$$

$$\Gamma^0{}_{33} = \frac{rC}{D}\sin^2\theta \tag{4.107}$$

$$\Gamma^1{}_{00} = \frac{1}{2D}(AA' + C\dot{A} - 2A\dot{C}) \tag{4.108}$$

$$\Gamma^1{}_{01} = \frac{1}{2D}(A'C + A\dot{B}) \tag{4.109}$$

$$\Gamma^1{}_{11} = \frac{1}{2D}(2CC' + AB' + C\dot{B}) \tag{4.110}$$

$$\Gamma^1{}_{22} = -\frac{rA}{D} \tag{4.111}$$

$$\Gamma^1{}_{33} = -\frac{rA}{D}\sin^2\theta \tag{4.112}$$

$$\Gamma^2{}_{12} = \Gamma^3{}_{13} = \frac{1}{r} \tag{4.113}$$

$$\Gamma^2{}_{33} = -\sin\theta\cos\theta \tag{4.114}$$

$$\Gamma^3{}_{23} = -\cot\theta. \tag{4.115}$$

Confirm that these results agree with those of Ex. 4.3 in the appropriate limiting case. For a shorter version of this problem, carry out this program for the first three affine connection coefficients.

4.5 Show that not all the affine connection coefficients vanish in Euclidean space in spherical coordinates. Thus, vanishing of the affine connection is not a criterion for a space to be "flat." If you worked Ex. 4.3 or Ex. 4.4, you can answer this question as a special case of those.

4.6 Propose the covariant derivative of A_μ with respect to a scalar τ, and verify that the result transforms as the appropriate first-order tensor.

4.7 Does $\Gamma^\lambda{}_{\mu\nu} = 0$ in a free-fall frame?

4.8 Show that Eq. (4.46) can also be derived by rewriting Eq. (4.43) for the reverse transformation $x'^\mu \to x^\mu$ and then solving for $\Gamma'^\lambda{}_{\mu\nu}$.

4.9 Show that

$$(\partial_\lambda T^\alpha)' = \frac{\partial x'^\alpha}{\partial x^\rho}\frac{\partial x^\sigma}{\partial x'^\lambda}(\partial_\sigma T^\rho) + \frac{\partial^2 x'^\alpha}{\partial x^\rho \partial x^\sigma}\frac{\partial x^\sigma}{\partial x'^\lambda}T^\rho. \tag{4.116}$$

4.10 Show that $DA^\lambda/D\tau$ transforms as a tensor of order 1.

4.11 Show that $A^\lambda{}_{;\mu}$ transforms as a mixed tensor of order 2.

4.12 (a) If you are familiar with the calculus of variations, show that, in an arbitrary Riemannian geometry, the free-fall equation

$$\frac{d^2x^\lambda}{d\tau^2} + \Gamma^\lambda{}_{\mu\nu}u^\mu u^\nu = 0 \tag{4.117}$$

is the Euler-Lagrange equation that emerges from requiring the proper time between two events a and b to be a maximum. In other words, require the functional

$$\Delta\tau = \int_a^b \sqrt{g_{\mu\nu}dx^\mu dx^\nu} = \int_a^b \sqrt{g_{\mu\nu}u^\mu u^\nu}\,d\tau \tag{4.118}$$

to be an extremal, with the Lagrangian

$$L = L(x^\rho, u^\rho) = \sqrt{g_{\mu\nu}u^\mu u^\nu} \tag{4.119}$$

and where $u^\mu = dx^\mu/d\tau$. This Lagrangian depends on the coordinates when the metric tensor does. The trajectory given by this result is called a "geodesic," the longest proper time between two fixed events in spacetime.
(b) A similar result holds in geometry, where the free-fall equation becomes the equation of a geodesic, the path of shortest distance on a surface or within a space between two points, if $d\tau$ gets replaced with path length ds. Show that a geodesic on the xy Euclidean plane is a straight line.
(c) Show that a geodesic on a spherical surface is a great circle (the intersection of a plane and a sphere, such that the plane passes through the sphere's center).

4.13 In the equation of the geodesic of Ex. 4.12, show that if τ is replaced with $w = \alpha\tau + \beta$, where α and β are constants, then the equation of the geodesic is satisfied with w as the independent variable. A parameter such as τ or w is called an "affine" parameter. Only linear transformations on the independent variable will preserve the equation of the geodesic; in other words, the geodesic equation is invariant only under an affine transformation.

4.14 In special and general relativity, which employ pseudo-Riemannian geometries, this notion of geodesic means that, for free-fall between two events, the proper time is a *maximum* (see Ex. 4.12), a "Fermat's principle" for free fall. Show that, in the Newtonian limit, the condition $\int d\tau = max.$ reduces to the functional of Hamilton's principle in classical mechanics, $\int (K - U)dt = min.$, where $K - U$ denotes the kinetic minus potential energy.

4.15 Show that

$$\Gamma^\mu{}_{\mu\lambda} = \frac{1}{\sqrt{|g|}}\partial_\lambda\sqrt{|g|}. \tag{4.120}$$

Notice that $\Gamma^\mu{}_{\mu\lambda} = \frac{1}{2}g^{\mu\rho}(\partial_\lambda g_{\rho\mu})$ and that $\partial_\lambda \mathrm{Tr}g = \partial_\lambda(g^{\mu\rho}g_{\rho\mu})$. Note also that for any matrix M, $\mathrm{Tr}[M^{-1}\partial_\lambda M] = \partial_\lambda \ln|M|$, where $|M|$ is the determinant of M. See Weinberg, pp. 106-107 for further hints.

4.16 Fill in the steps for the derivation of the Laplacian made with covariant derivatives, Eq. (4.92).

Chapter 5

Curvature

5.1 What Is Curvature?

In calculus we learned that for a function $y = f(x)$, dy/dx measures the slope of the tangent line, and d^2y/dx^2 tells about curvature. For instance, if the second derivative is positive at x, then the curve there opens upward. More quantitatively, as demonstrated in Ex. 5.3, the local curvature κ (viz., the inverse radius of the osculating circle, where the Latin *osculari* means "to kiss") of $y = f(x)$ is

$$\kappa = \frac{\left|\frac{d^2y}{dx^2}\right|}{\left[1 + \left(\frac{dy}{dx}\right)^2\right]^{\frac{3}{2}}}. \tag{5.1}$$

An obvious test case would be a circle of radius R centered on the origin, for which $y = \pm\sqrt{R^2 - x^2}$. From Eq. (5.1) it follows that $\kappa = \frac{1}{R}$, as expected.

A curve is a one-dimensional space according to a point sliding along it. But to measure the curvature at a point on $y = f(x)$ through the method used to derive κ, the curve had to be embedded in the two-dimensional xy plane. Similarly, the curvature of the surface of a globe or a saddle is easily seen when observed from the perspective of three-dimensional space. Pronouncements on the curvature of a space of n dimensions, which depend on the space being embedded in a space of $n + 1$ dimensions, are called *extrinsic* measures of curvature.

To a little flat bug whose entire world consists of the surface of the globe or saddle, who has no access to the third dimension off the surface, any local patch of area in its immediate vicinity appears flat and could be mapped with a local Euclidean xy coordinate system. How would our bug discover that its world is not *globally* Euclidean? Since our dimuntitive friend cannot step off the surface, it must make measurements within the space itself. Such internal measures of the curvature of a space are said to be *intrinsic*, probing the "inner properties" of the space.

For example, our bug could stake out three locations and connect them with survey lines that cover the shortest distance on the surface between each pair of points. Such "shortest distance" trajectories are called *geodesics* (recall Ex. 4.12). Once the goedesics are surveyed, our bug measures the sum of the interior angles of the triangle. If the space is flat across the area outlined by the three points, these interior angles add up to 180 degrees. If the sum of the interior angles is greater than 180 degrees, the surface has positive curvature (e.g., the surface of the globe). If the sum of interior angles is less than 180 degrees, the surface has negative curvature, like a hyperbolic surface.

The bug could also mark out two lines that are locally parallel. If the space is everywhere flat, those parallel lines will never meet, as Euclid's fifth postulate asserts. If the locally parallel lines are found to intersect, although each one follows a geodesic, the geometry is spherical (on a globe, longitudinal lines are parallel as they cross the equator, but meet at the poles). If the geodesics diverge, then the geometry is hyperbolic.

Suppose the three-dimensional universe that birds and galaxies fly through is actually the surface of a sphere in a four-dimensional Euclidean space (this was Albert Einstein's first stab at cosmological applications of general relativity, which he published in 1917, and started a conversation that led over the next decade, through the work of others, to big bang cosmology). Spacetime would then have five dimensions. But we have no access to that fourth spatial dimension. Even if it exists, we are not able to travel into it, and would find ourselves, in three dimensions, in the same predicament as the bugs in their two-dimensional surface. From *within* a space, the geometry can *only* be probed intrinsically.

To illustrate why intrinsic measures of curvature are more fundamental than extrinsic measurements, take a sheet of paper, manifestly a plane. Draw parallel lines on it; draw a triangle and measure the interior angles; measure the distance on the paper between any two points. Then roll the paper into a cylinder. The angles and distances are unchanged, and lines that were parallel on the flat sheet are still parallel on the cylindrical roll. The intrinsic geometry is the same in the cylinder as in the Euclidean plane. What has changed is the topology: Our bug can walk around the cylinder, never change direction, and return to the starting place.

Topology is important, but for physical measurements we need distances; we need the space to be a *metric space*. The metric on the flat sheet and that on the cylinder are the same; the distance between any two dots as measured *on the paper itself* is unchanged by rolling the plane into a cylinder. The cylinder is as flat as the plane in the sense that the Pythagorean theorem holds on both when the surface's inner properties are surveyed. In contrast, no *single* sheet of paper can accurately map the entirety of the Earth's surface; something always gets distorted.

Another example occurs in the Schwarzschild metric of Eq. (3.53), which describes the spacetime around a star. The r-coordinate is *not* the radial distance from the origin; rather, it is defined as the circumference, divided by 2π, of a spherical shell centered on the origin. Ex. 3.10 described how, for two concentric spherical shells with r-coordinates r_1 and $r_2 > r_1$, the distance between them is given, not by $\int_{r_1}^{r_2} dr = r_2 - r_1$ as would be the case in flat Euclidean space, but instead by

$$\int_{r_1}^{r_2} \frac{dr}{1 - \frac{2GM}{rc^2}}. \tag{5.2}$$

Measuring the radial distance with a meterstick, and getting something different from $r_2 - r_1$, offers a tipoff that geometry around the star is curved.

Another intrinsic way to determine whether a space is curved comes from the notion of "parallel transport" of a vector. To move the Euclidean vector **A** from point P to point Q, some path $\mathcal{C}$ is chosen that connects P to Q. Suppose that points on the path from P to Q along contour $\mathcal{C}$ are distinguished by a parameter s, such as the distance from P. Let A^μ label the components of **A**. Transport **A** from s to $s + ds$. For **A** to be parallel-transported, we require, as the definition of parallel transport, that its components remain the same at any infinitesimal step in the transport along $\mathcal{C}$,

$$A^\mu(s + ds) - A^\mu(s) = 0, \tag{5.3}$$

or

$$\frac{dA^\mu}{ds} = 0, \tag{5.4}$$

which holds locally since any space is locally flat. If the space is *globally* flat, then Eq. (5.3) holds everywhere along any closed path $\mathcal{C}$, and thus for every component of **A**,

$$\oint_{\mathcal{C}} dA^\mu = 0. \tag{5.5}$$

Try it with a closed contour sketched on a sheet of paper. Draw any vector **A** whose tail begins at any point on $\mathcal{C}$. At an infinitesimally nearby point on $\mathcal{C}$, draw a new vector having the same magnitude as **A**, and keep the new vector parallel to its predecessor. Continue this process all the way around the closed path. The final vector coincides with the initial one, and thus the closed path line integral vanishes. In three-dimensional Euclidean space any vector **A** can be parallel-transported through space, along any contour, while keeping its direction and magnitude constant.

Now try the same procedure on the surface of a sphere. Examine a globe representing the Earth. Start at a point on the equator, say, where the Pacific Ocean meets the west coast of South America, on the coast of Ecuador. Let your pencil represent a vector. Point your pencil-vector due east, parallel to the equator. Slide the vector north along the line of constant longitude. At each incremental step on the journey keep the new vector parallel to its infinitesimally nearby predecessor. Your trajectory passes near Havana, New York City, Prince Charles Island in the North Hudson Bay, and on to the North Pole. As you arrive at the North Pole, note the direction in which your vector points. Now move along a line of another longitude, such as the longitude of 0 degrees that passes through Greenwich, England. As you leave the North Pole, keep each infinitesimally translated vector as parallel as you can to the one preceding it. This trajectory passes through Greenwich, goes near Madrid and across western Africa, arriving at the equator south of Ghana. Keeping your vector parallel to the one immediately preceding it, slide it west along the equator until you return to the Pacific beach at Ecuador.

Notice what has happened: the initial vector pointed due east, but upon returning to the starting place the final vector points southeast! Therefore, the difference between the inital and final vectors is not zero, and the line integral does not vanish: the hallmark of a curved space. We take it as an operational definition to say that a space is curved if and only if

$$\oint_{\mathcal{C}} dA^{\mu} \neq 0 \tag{5.6}$$

for any closed path $\mathcal{C}$ in the space and for arbitrary A^{μ}.

This definition does not tell us *what* $\oint dA^{\mu}$ happens to be, should it not equal zero. Ah, but our subject is tensors, so when using Eq. (5.6) as a criterion for curved space, we should take into account the covariant derivative. Perhaps *it* will tell us what the closed path line integral happens to be when it does not vanish. Since the covariant derivative algorithm has "$\partial \to D = \partial + \Gamma - term$," the affine connection coefficients may have something to say about this.

Because curvature is related to the second derivative, and the derivatives of tensors must be covariant derivatives, in Section 5.2 we consider second covariant derivatives. In Section 5.3 we return to the closed path integral Eq. (5.6), informed by covariant derivatives, and find how it is related to second covariant derivatives and curvature.

5.2 The Riemann Tensor

Since the differential equations of physics are frequently of second order, with some curiosity we must see what arises in *second* covariant derivatives. That forms the subject of this section. In the next section we will see what this has to do with curvature.

The covariant first derivative of a vector component with respect to a coordinate, it will be recalled, is defined according to

$$D_{\nu}A^{\mu} \equiv \partial_{\nu}A^{\mu} + \Gamma^{\mu}{}_{\rho\nu}A^{\rho}, \tag{5.7}$$

which makes $D_{\nu}A^{\mu}$ a second-order tensor. Consider the second covariant derivative, $D_{\rho}(D_{\nu}A^{\mu}) \equiv (D_{\nu}A^{\mu})_{;\rho} \equiv A^{\mu}{}_{;\nu;\rho}$. By the rules of evaluating the covariant derivative of a second-order tensor, we have

$$\begin{aligned}(D_{\nu}A^{\mu})_{;\rho} &= (D_{\nu}A^{\mu})_{,\rho} + \Gamma^{\mu}{}_{\sigma\rho}(D_{\nu}A^{\sigma}) - \Gamma^{\sigma}{}_{\nu\rho}(D_{\sigma}A^{\mu}) \\ &= (A^{\mu}{}_{,\nu} + \Gamma^{\mu}{}_{\sigma\nu}A^{\sigma})_{,\rho} + \Gamma^{\mu}{}_{\sigma\rho}(A^{\sigma}{}_{,\nu} + \Gamma^{\sigma}{}_{\tau\nu}A^{\tau}) - \Gamma^{\sigma}{}_{\nu\rho}(A^{\mu}{}_{,\sigma} + \Gamma^{\mu}{}_{\tau\sigma}A^{\tau}).\end{aligned}$$

Recall from calculus that when evaluating the second derivative of a function of two independent variables, $\varphi = \varphi(x,y)$, the order of differentiation will not matter,

$$\frac{\partial^2\varphi}{\partial x\partial y} = \frac{\partial^2\varphi}{\partial y\partial x}, \tag{5.8}$$

provided that both of the second derivatives exist and are continuous at the point evaluated. In other words, for non-pathological functions the commutator of the derivatives vanishes, where the "commutator" $[A,B]$ of two quantities A and B means

$$[A,B] \equiv AB - BA. \tag{5.9}$$

In terms of commutators, Eq. (5.8) may be written

$$\left[\frac{\partial}{\partial x},\frac{\partial}{\partial y}\right]\varphi = \left[\frac{\partial}{\partial x}\frac{\partial}{\partial y} - \frac{\partial}{\partial y}\frac{\partial}{\partial x}\right]\varphi = 0. \tag{5.10}$$

Does so simple a conclusion hold for second covariant derivatives? Let us see.

First, reverse the order of taking the covariant derivatives, by switching ρ and ν in $(D_\nu A^\mu)_{;\rho}$ to construct

$$(D_\rho A^\mu)_{;\nu} = (A^\mu{}_{,\rho} + \Gamma^\mu{}_{\sigma\rho}A^\sigma)_{,\nu} + \Gamma^\mu{}_{\sigma\nu}(A^\sigma{}_{,\rho} + \Gamma^\sigma{}_{\tau\rho}A^\tau) - \Gamma^\sigma{}_{\rho\nu}(A^\mu{}_{,\sigma} + \Gamma^\mu{}_{\tau\sigma}A^\tau). \tag{5.11}$$

Now comes the test. Evaluate the difference between these second derivative orderings. Invoking the symmetry of the affine connection symbols, $\Gamma^\lambda{}_{\mu\nu} = \Gamma^\lambda{}_{\nu\mu}$, we find

$$(D_\rho D_\nu - D_\nu D_\rho)A^\mu = R^\mu{}_{\nu\tau\rho}A^\tau, \tag{5.12}$$

where

$$R^\mu{}_{\nu\tau\rho} \equiv \partial_\rho\Gamma^\mu{}_{\nu\tau} - \partial_\nu\Gamma^\mu{}_{\rho\tau} + \Gamma^\mu{}_{\rho\sigma}\Gamma^\sigma{}_{\nu\tau} - \Gamma^\mu{}_{\nu\sigma}\Gamma^\sigma{}_{\rho\tau}. \tag{5.13}$$

This complicated combination of derivatives and products of affine connection coefficients denoted $R^\mu{}_{\nu\tau\rho}$ constitute the components of the "Riemann tensor." As you may show (if you have great patience), the Riemann tensor respects the transformation rule for a fourth-order tensor because the parts of the affine connection transformation which spoil the tensor character cancel out.

With its four indices, in three-dimensional space the Riemann tensor has $3^4 = 81$ components. In four-dimensional spacetime it carries $4^4 = 256$ components. However, symmetries exist that reduce the number of independent components. For instance, of the 256 components in 4-dimensional spacetime, only 20 are independent. The symmetries are conveniently displayed by shifting all the indices downstairs with the assistance of the metric tensor,

$$R_{\mu\nu\rho\sigma} = g_{\mu\tau}R^\tau{}_{\nu\rho\sigma}. \tag{5.14}$$

The Riemann tensor is antisymmetric under the exchange of its first two indices,

$$R_{\alpha\beta\mu\nu} = -R_{\beta\alpha\mu\nu}, \tag{5.15}$$

and under the exchange of its last two indices,

$$R_{\alpha\beta\mu\nu} = -R_{\alpha\beta\nu\mu}. \tag{5.16}$$

It is symmetric when swapping the first pair of indices with the last pair of indices:

$$R_{\alpha\beta\mu\nu} = R_{\mu\nu\alpha\beta}. \tag{5.17}$$

The Riemann tensor also respects a cyclic property among its last three components,

$$R_{\alpha\beta\mu\nu} + R_{\alpha\mu\nu\beta} + R_{\alpha\nu\beta\mu} = 0, \tag{5.18}$$

and the differential "Bianchi identities" that keep the first two indices intact but permute the others:

$$R_{\alpha\beta\mu\nu;\sigma} + R_{\alpha\beta\nu\sigma;\mu} + R_{\alpha\beta\sigma\mu;\nu} = 0. \tag{5.19}$$

When two of the Riemann tensor indices are contracted, the second-order tensor that results is called the "Ricci tensor" $R_{\mu\nu}$,

$$\begin{aligned} R_{\mu\nu} &\equiv R^{\rho}{}_{\mu\rho\nu} \\ &= \partial_\nu \Gamma^{\rho}{}_{\mu\rho} - \partial_\rho \Gamma^{\rho}{}_{\mu\nu} + \Gamma^{\lambda}{}_{\mu\rho}\Gamma^{\rho}{}_{\lambda\nu} - \Gamma^{\lambda}{}_{\mu\nu}\Gamma^{\rho}{}_{\lambda\rho}, \end{aligned}$$

which is symmetric,

$$R_{\mu\nu} = R_{\nu\mu}. \tag{5.20}$$

Raising one of the indices on the Ricci tensor presents it in mixed form, $R^{\mu}{}_{\rho} = g^{\mu\nu}R_{\nu\rho}$, which can be contracted into the "Riemann scalar" R,

$$R \equiv R^{\mu}{}_{\mu}. \tag{5.21}$$

The Ricci tensor and the scalar R offer up another form of Bianchi identity,

$$(R^{\mu\nu} - \frac{1}{2}g^{\mu\nu}R)_{;\mu} = 0. \tag{5.22}$$

In a schematic "cartoon equation" that overlooks subscripts and superscripts and permutations, since $\Gamma \sim g\partial g$, we can say that the Riemann tensor components have the structure

$$\begin{aligned} R &\sim \partial\Gamma + \Gamma\Gamma \\ &\sim g\partial^2 g + (\partial g)^2 + (g\partial g)^2. \end{aligned}$$

Although the Riemann tensor is quadratic in the first derivatives of metric tensor coefficients, it is the *only* tensor linear in the second derivative of the metric tensor, as will be demonstrated in Section 5.4. This feature makes the Riemann tensor important for applications, as we shall see.

5.3 Measuring Curvature

Now we return to the criterion for a space to be curved–the nonvanishing of the closed path integral $\oint_{\mathcal{C}} dA^{\mu}$ for every closed path $\mathcal{C}$. If the integral is not zero, what is it? The answer to that question should give some measure of curvature.

This closed path line integral criterion began from the observation that, for a vector undergoing parallel transport from one point on $\mathcal{C}$ to another point on $\mathcal{C}$ infinitesimally close by, with s a parameter along $\mathcal{C}$,

$$\frac{dA^{\mu}}{ds} = 0. \tag{5.23}$$

As the component of a vector field, A^{μ} depends on the coordinates, so that by the chain rule,

$$\begin{aligned} \frac{dA^{\mu}}{ds} &= \frac{\partial A^{\mu}}{\partial x^{\nu}} \frac{dx^{\nu}}{ds} \\ &= \frac{\partial A^{\mu}}{\partial x^{\nu}} u^{\nu}, \end{aligned}$$

where

$$u^{\nu} \equiv \frac{dx^{\nu}}{ds}. \tag{5.24}$$

But for this to be a tensor equation, the covariant derivative must be taken into account. Therefore, the tensor equation of parallel transport says

$$\frac{DA^{\mu}}{Ds} = 0. \tag{5.25}$$

Writing out the covariant derivative gives

$$\frac{dA^{\mu}}{ds} + \Gamma^{\mu}{}_{\nu\rho} u^{\nu} A^{\rho} = 0. \tag{5.26}$$

Notice that if $A^{\nu} = u^{\nu}$, then Eq. (5.26) is the equation of a geodesic (recall Ex. 4.12). Thus, an alternative definition of parallel transport says that a tangent vector undergoing parallel transport follows a geodesic.

Now the condition for parallel transport, Eq. (5.23), generalized to the covariant derivative, may be written

$$\frac{dA^{\mu}}{ds} = -\Gamma^{\mu}{}_{\nu\rho} u^{\nu} A^{\rho}. \tag{5.27}$$

We could try to integrate this at once and confront

$$\oint_{\mathcal{C}} dA^{\mu} = -\oint_{\mathcal{C}} \Gamma^{\mu}{}_{\nu\rho} u^{\nu} A^{\rho} ds. \tag{5.28}$$

By the criterion for a space to be curved, $\oint_{\mathcal{C}} dA^{\mu} \neq 0$, the integral on the right-hand side of Eq. (5.28) would not vanish. However, the right-hand side of Eq. (5.28) is more complicated than necessary for addressing our problem. Return to Eq. (5.26) and again invoke the chain rule. Eq. (5.26) becomes

$$u^{\nu}\left[\frac{\partial A^{\mu}}{\partial x^{\nu}}+\Gamma^{\mu}{}_{\nu\rho}A^{\rho}\right]=0. \tag{5.29}$$

From Eq. (5.29), the rate of change of a vector component with respect to a coordinate, when undergoing parallel transport, satisfies the equation

$$\frac{\partial A^{\mu}}{\partial x^{\nu}}=-\Gamma^{\mu}{}_{\nu\rho}A^{\rho}. \tag{5.30}$$

Given initial conditions, one can find the effect on the vector component, when moving from point P to point Q, by integrating Eq. (5.30),

$$A^{\mu}(Q)-A^{\mu}(P)=-\int_{P}^{Q}\Gamma^{\mu}{}_{(\nu)\rho}A^{\rho}dx^{(\nu)}, \tag{5.31}$$

where I have put parentheses around ν to remind us that, in this integral, we do not sum over the index ν.

Now return to the working definition of curvature: making points Q and P the same point *after* traversing some closed path, we can say that if any closed path C exists for which

$$\oint_{C}dA^{\mu}=-\oint_{C}\Gamma^{\mu}{}_{(\nu)\rho}A^{\rho}dx^{(\nu)}\neq 0, \tag{5.32}$$

then the space is said to be curved.

Eq. (5.32) also tells us *what* the line integral of dA^{μ} around a closed path actually is, when it does not equal zero. For that purpose, please refer to Fig. 5.1. The figure shows a contour made of four different sections, from point 1 to point 2, then from point 2 to point 3. From there we go to point 4, and then from point 4 back to point 1. Each segment follows either a contour of constant x^1, at $x^1=a$ or $a+\Delta a$; or constant $x^2=b$ or $b+\Delta b$. Let I, J, K, and L denote the line integrals to be computed according to Eq. (5.32). Traversing the closed path one section at a time, we write each segment of the closed path integral using Eq. (5.31),

$$\begin{aligned}
A^{\mu}(2) &= A^{\mu}(1_{initial})+I,\\
A^{\mu}(3) &= A^{\mu}(2)+J,\\
A^{\mu}(4) &= A^{\mu}(3)-K,\\
A^{\mu}(1_{final}) &= A^{\mu}(4)-L,
\end{aligned}$$

where the minus signs are used with K and L because the line integrals will be evaluated from a to $a+\Delta a$ and from b to $b+\Delta b$, but those segments of the contour are traversed in the opposite directions. For the entire contour, these four segments combine to give

$$\begin{aligned}
\oint dA^{\mu} &= A^{\mu}(1_{final})-A^{\mu}(1_{initial})\\
&= (I-K)+(J-L).
\end{aligned}$$

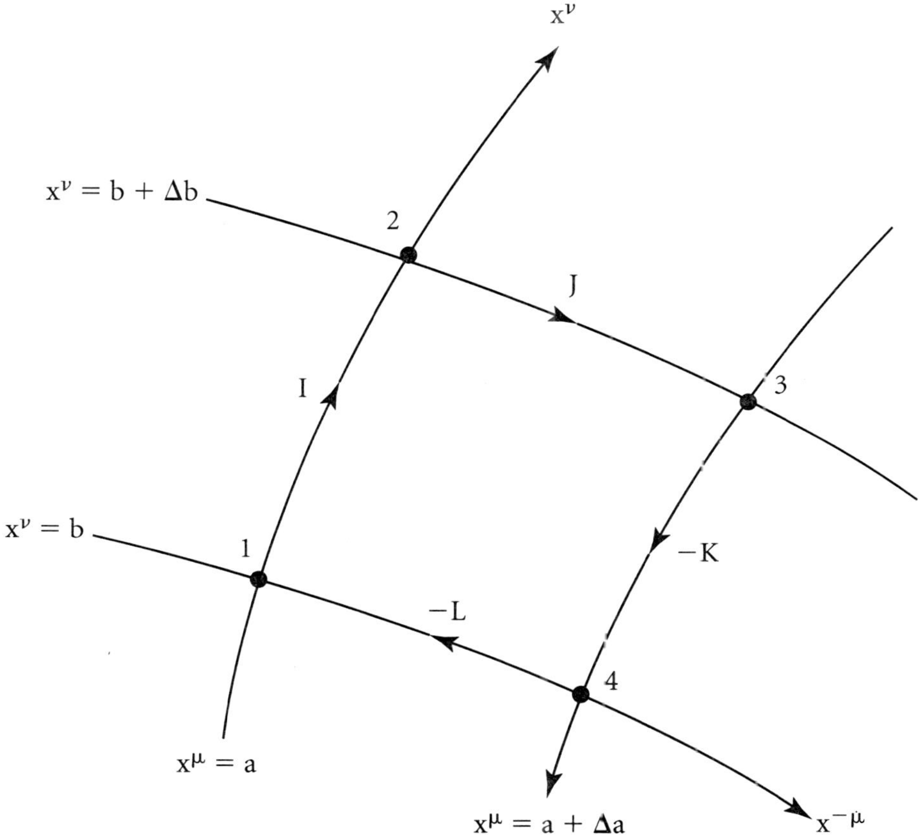

Figure 5.1: *The closed path used in evaluating $\oint dA^\mu$. The surface bounded by an arbitrary contour $\mathcal{C}$ can be tiled with an array of such infinitesimal contours that the only contributions that do not automatically cancel out come from the perimeter.*

By virtue of Eq. (5.31), the line integrals for each segment are

$$I = -\int_b^{b+\Delta b} \left[\Gamma^\mu{}_{2\rho} A^\rho\right]_{x^1=a} dx^2, \tag{5.33}$$

$$J = -\int_a^{a+\Delta a} \left[\Gamma^\mu{}_{1\rho} A^\rho\right]_{x^2=b+\Delta b} dx^1, \tag{5.34}$$

$$K = -\int_b^{b+\Delta b} \left[\Gamma^\mu{}_{2\rho} A^\rho\right]_{x^1=a+\Delta a} dx^2, \tag{5.35}$$

and

$$L = -\int_a^{a+\Delta a} \left[\Gamma^\mu{}_{1\rho} A^\rho\right]_{x^2=b} dx^1. \tag{5.36}$$

Putting in these expressions for the line integrals, we find

$$\begin{aligned}
\oint dA^\mu &= -\int_b^{b+\Delta b} \left[(\Gamma^\mu{}_{2\rho} A^\rho)_{x^1=a} - (\Gamma^\mu{}_{2\rho} A^\rho)_{x^1=a+\Delta a} \right] dx^2 \\
&\quad - \int_a^{a+\Delta a} \left[(\Gamma^\mu{}_{1\rho} A^\rho)_{x^2=b+\Delta b} - (\Gamma^\mu{}_{2\rho} A^\rho)_{x^2=b} \right] dx^1 \\
&\approx \int_b^{b+\Delta b} \Delta a \frac{\partial}{\partial x^1} (\Gamma^\mu{}_{2\rho} A^\rho)\, dx^2 - \int_a^{a+\Delta a} \Delta b \frac{\partial}{\partial x^2} (\Gamma^\mu{}_{1\rho} A^\rho)\, dx^1 \\
&\approx \Delta a \Delta b \left[\partial_1 (\Gamma^\mu{}_{2\rho} A^\rho) - \partial_2 (\Gamma^\mu{}_{1\rho} A^\rho) \right].
\end{aligned}$$

Evaluating the derivatives of products and with the help of Eq. (5.30) and some relabeling of repeated indices, this becomes

$$\oint dA^\mu \approx \Delta a \Delta b \left[\frac{\partial \Gamma^\mu{}_{2\rho}}{\partial x^1} - \frac{\partial \Gamma^\mu{}_{1\rho}}{\partial x^2} + \Gamma^\mu{}_{1\sigma} \Gamma^\sigma{}_{2\rho} - \Gamma^\mu{}_{2\sigma} \Gamma^\sigma{}_{1\rho} \right] A^\rho. \tag{5.37}$$

Since the labels x^1 and x^2 are arbitrary, we can replace them with x^τ and x^λ, respectively, and obtain at last

$$\oint_{(\lambda,\tau)} dA^\mu \approx \Delta a \Delta b \left[\frac{\partial \Gamma^\mu{}_{\lambda\rho}}{\partial x^\tau} - \frac{\partial \Gamma^\mu{}_{\tau\rho}}{\partial x^\lambda} + \Gamma^\mu{}_{\tau\sigma} \Gamma^\sigma{}_{\lambda\rho} - \Gamma^\mu{}_{\lambda\sigma} \Gamma^\sigma{}_{\tau\rho} \right] A^\rho, \tag{5.38}$$

where the (λ, τ) notation on the integral reminds us that the integral is evaluated along segments of fixed x^λ or x^τ. The quantity in the square brackets will be recognized from Section 5.2 as a component of the Riemann tensor,

$$R^\mu{}_{\lambda\tau\rho} \equiv \frac{\partial \Gamma^\mu{}_{\lambda\rho}}{\partial x^\tau} - \frac{\partial \Gamma^\mu{}_{\tau\rho}}{\partial x^\lambda} + \Gamma^\mu{}_{\tau\sigma} \Gamma^\sigma{}_{\lambda\rho} - \Gamma^\mu{}_{\lambda\sigma} \Gamma^\sigma{}_{\tau\rho}. \tag{5.39}$$

In terms of the Riemann tensor, our contour integral has become

$$\oint_{(\lambda,\tau)} dA^\mu \approx (\Delta a \Delta b) R^\mu{}_{\lambda\tau\rho} A^\rho. \tag{5.40}$$

For arbitrary $\Delta a \Delta b$ and arbitrary A^ρ, the closed path integral shows zero curvature if and only if all the components of the Riemann tensor vanish. Thus, the $R^\mu{}_{\lambda\tau\rho}$ are justfiably called the components of the Riemann *curvature* tensor. We study its uniqueness in the next section.

5.4 Linearity in the Second Derivative

Albert Einstein constructed general relativity by applying the principle of general covariance to Poisson's equation, whose solution gives the Newtonian gravitational potential. Poisson's equation happens to be linear in the second derivative of the potential, and from the Newtonian limit of the free-fall (or geodesic) equation, the g_{00} component of the metric tensor yields the Newtonian gravitational potential (recall $g_{00} \approx 1 + 2\phi$ in Section 4.3). To make general relativity no more complicated than necessary, Einstein hypothesized that his field equations should also be linear in the second derivative of the metric tensor. Since the Riemann tensor offers the only tensor linear in the second derivative of $g_{\mu\nu}$, Einstein turned to it. This claim of linearity can be demonstrated. We proceed by constructing a tensor that is linear in the second derivative of the metric tensor, and show that it equals the Riemann tensor. Let us proceed in six steps.

First, begin with the "alternative" transformation of the affine connection, Eq. (4.46),

$$\Gamma'^{\lambda}{}_{\mu\nu} = \frac{\partial x'^{\lambda}}{\partial x^{\rho}} \frac{\partial x^{\tau}}{\partial x'^{\mu}} \frac{\partial x^{\sigma}}{\partial x'^{\nu}} \Gamma^{\rho}{}_{\tau\sigma} - \frac{\partial x^{\rho}}{\partial x'^{\nu}} \frac{\partial x^{\sigma}}{\partial x'^{\mu}} \frac{\partial^2 x'^{\lambda}}{\partial x^{\rho} \partial x^{\sigma}}. \tag{5.41}$$

Because of an avalanche of partial derivatives that will soon be upon us, let us introduce in this section the shorthand abbreviations

$$\Lambda^{\lambda'}_{\rho} \equiv \frac{\partial x'^{\lambda}}{\partial x^{\rho}} \tag{5.42}$$

and

$$\Lambda^{\lambda}_{\rho'} \equiv \frac{\partial x^{\lambda}}{\partial x'^{\rho}}. \tag{5.43}$$

Second derivatives can be denoted

$$\Lambda^{\lambda'}_{\mu\eta'} \equiv \frac{\partial \Lambda^{\lambda'}_{\mu}}{\partial x'^{\eta}} \equiv \frac{\partial^2 x'^{\lambda}}{\partial x^{\mu} \partial x'^{\eta}}. \tag{5.44}$$

For the third derivatives a four-index symbol $\Lambda^{\lambda}_{\rho\sigma\alpha}$ could be defined along the same lines, but the third derivatives will be conspicuous in what follows. In this notation, the transformation just cited for the affine connection reads

$$\Gamma'^{\lambda}{}_{\mu\nu} = \Lambda^{\lambda'}_{\rho} \Lambda^{\tau}_{\mu'} \Lambda^{\sigma}_{\nu'} \Gamma^{\rho}{}_{\tau\sigma} - \Lambda^{\rho}_{\nu'} \Lambda^{\sigma}_{\mu'} \Lambda^{\lambda'}_{\rho\sigma}. \tag{5.45}$$

Second, because the Riemann curvature tensor features first derivatives of the affine connection coefficients, we must consider their transformation. Upon differentiating Eq. (5.45),

$$\begin{aligned} \frac{\partial \Gamma'^{\lambda}{}_{\mu\nu}}{\partial x'^{\eta}} = & \left(\Lambda^{\lambda'}_{\rho\eta'} \Lambda^{\tau}_{\mu'} \Lambda^{\sigma}_{\nu'} + \Lambda^{\lambda'}_{\rho} \Lambda^{\tau}_{\mu'\eta'} \Lambda^{\sigma}_{\nu'} + \Lambda^{\lambda'}_{\rho} \Lambda^{\tau}_{\mu'} \Lambda^{\sigma}_{\nu'\eta'} \right) \Gamma^{\rho}{}_{\tau\sigma} + \Lambda^{\lambda'}_{\rho} \Lambda^{\tau}_{\mu'} \Lambda^{\sigma}_{\nu'} \frac{\partial \Gamma^{\rho}{}_{\tau\sigma}}{\partial x'^{\eta}} \\ & - \left(\Lambda^{\rho}_{\nu'\mu'} \Lambda^{\sigma}_{\mu'} + \Lambda^{\rho}_{\nu'} \Lambda^{\sigma}_{\mu'\eta'} \right) \Lambda^{\lambda'}_{\rho\sigma} - \Lambda^{\rho}_{\nu'} \Lambda^{\sigma}_{\mu'} \frac{\partial^3 x'^{\lambda}}{\partial x^{\rho} \partial x^{\sigma} \partial x^{\alpha}} \Lambda^{\alpha}_{\eta'}, \end{aligned}$$

where the chain rule was used in the last term.

Third, evaluate this expression at an event P for which all the second derivatives of *primed* coordinates with respect to the *unprimed* coordinates vanish,

$$\Lambda^{\lambda'}_{\mu\nu}|_P = \frac{\partial^2 x'^{\lambda}}{\partial x^{\mu} \partial x^{\nu}}|_P = 0, \tag{5.46}$$

which means (because of the original definition of the affine connection coefficients) that

$$\Gamma^{\rho}{}_{\mu\nu}|_P = \frac{\partial x^{\rho}}{\partial x'^{\lambda}}\frac{\partial^2 x'^{\lambda}}{\partial x^{\mu}\partial x^{\nu}}|_P = 0. \tag{5.47}$$

The set of such points P is not the null set: in the gravitational context, such an event P occurs in a *locally* inertial free-float frame, and in a geometry context, *locally* the space can be taken to be Euclidean.

Evaluated in a local patch of spacetime near event P, in our long expression for $\partial\Gamma'^{\lambda}{}_{\mu\nu}/dx'^{\eta}$ above, the first and third terms vanish, leaving

$$\frac{\partial\Gamma'^{\lambda}{}_{\mu\nu}}{\partial x'^{\eta}}|_P = \Lambda^{\lambda'}_{\rho}\Lambda^{\tau}_{\mu'}\Lambda^{\sigma}_{\nu'}\frac{\partial\Gamma^{\rho}{}_{\tau\sigma}}{\partial x'^{\eta}} - \Lambda^{\rho}_{\nu'}\Lambda^{\sigma}_{\mu'}\frac{\partial^3 x'^{\lambda}}{\partial x^{\rho}\partial x^{\sigma}\partial x^{\alpha}}\Lambda^{\alpha}_{\eta'}. \tag{5.48}$$

With the assistance of the chain rule, the first term on the right-hand side can be rewritten in terms of the derivative of the Γ-coefficient with respect to an x-coordinate rather than with respect to an x'-coordinate:

$$\frac{\partial\Gamma'^{\lambda}{}_{\mu\nu}}{\partial x'^{\eta}}|_P = \Lambda^{\lambda'}_{\rho}\Lambda^{\tau}_{\mu'}\Lambda^{\sigma}_{\nu'}\Lambda^{\alpha}_{\eta'}\frac{\partial\Gamma^{\rho}{}_{\tau\sigma}}{\partial x^{\alpha}} - \Lambda^{\rho}_{\nu'}\Lambda^{\sigma}_{\mu'}\Lambda^{\alpha}_{\eta'}\frac{\partial^3 x'^{\lambda}}{\partial x^{\rho}\partial x^{\sigma}\partial x^{\alpha}}. \tag{5.49}$$

Fourth, we note that $\Gamma^{\lambda}{}_{\mu\nu} = \Gamma^{\lambda}{}_{\nu\mu}$. Define the quantity

$$T^{\lambda}{}_{\mu\nu\eta} \equiv \partial_{\eta}\Gamma^{\lambda}{}_{\mu\nu} - \partial_{\nu}\Gamma^{\lambda}{}_{\mu\eta}, \tag{5.50}$$

which coinicides with the first two terms of the Riemann curvature tensor, the terms that are linear in the second derivative of the metric tensor. With the aid of Eq. (5.49), the transformation behavior of $T^{\lambda}{}_{\mu\nu\eta}$ can be investigated:

$$\begin{aligned}
T'^{\lambda}{}_{\mu\nu\eta} &= \partial_{\eta'}\Gamma'^{\lambda}{}_{\mu\nu} - \partial_{\nu'}\Gamma'^{\lambda}{}_{\mu\eta} \\
&= \Lambda^{\lambda'}_{\rho}\Lambda^{\tau}_{\mu'}\Lambda^{\sigma}_{\nu'}\Lambda^{\alpha}_{\eta'}(\partial_{\alpha}\Gamma^{\rho}{}_{\tau\sigma}) - \Lambda^{\rho}_{\nu'}\Lambda^{\sigma}_{\mu'}\Lambda^{\alpha}_{\eta'}\frac{\partial^3 x'^{\lambda}}{\partial x^{\rho}\partial x^{\sigma}\partial x^{\alpha}} \\
&- \Lambda^{\lambda'}_{\rho}\Lambda^{\tau}_{\mu'}\Lambda^{\sigma}_{\eta'}\Lambda^{\alpha}_{\nu'}(\partial_{\alpha}\Gamma^{\rho}{}_{\tau\sigma}) + \Lambda^{\rho}_{\eta'}\Lambda^{\sigma}_{\mu'}\Lambda^{\alpha}_{\nu'}\frac{\partial^3 x'^{\lambda}}{\partial x^{\rho}\partial x^{\sigma}\partial x^{\alpha}}.
\end{aligned}$$

In the fourth term, interchange the ρ and α, which can be done since they are summed out. Then the second and fourth terms (the terms with the third derivatives) are seen to cancel. To group the surviving first and third terms usefully, the Λ coefficients can be factored out if, in the third term, σ and α are interchanged, leaving us with

$$\begin{aligned}
T'^{\lambda}{}_{\mu\nu\eta} &= \Lambda^{\lambda'}_{\rho}\Lambda^{\tau}_{\mu'}\Lambda^{\alpha}_{\nu'}\Lambda^{\sigma}_{\eta'}\left[\frac{\partial\Gamma^{\rho}{}_{\tau\sigma}}{\partial x^{\alpha}} - \frac{\partial\Gamma^{\rho}{}_{\tau\alpha}}{\partial x^{\sigma}}\right] \\
&= \frac{\partial x'^{\lambda}}{\partial x^{\rho}}\frac{\partial x^{\tau}}{\partial x'^{\mu}}\frac{\partial x^{\alpha}}{\partial x'^{\nu}}\frac{\partial x^{\sigma}}{\partial x'^{\eta}}T^{\rho}{}_{\tau\sigma\alpha},
\end{aligned}$$

which is the transformation rule for a fourth-order tensor. Thus, $T^{\lambda}{}_{\mu\nu\eta}$ is a tensor.

Fifth, we notice, in the frame and at the event P where

$$\Gamma^{\lambda}{}_{\mu\nu}|_P = 0 \tag{5.51}$$

because

$$\frac{\partial^2 x'^{\lambda}}{\partial x^{\mu}\partial x^{\nu}}|_P = 0, \tag{5.52}$$

that

$$T^{\lambda}{}_{\mu\nu\eta}|_P = R^{\lambda}{}_{\mu\nu\eta}|_P. \tag{5.53}$$

By the principle of general covariance,

$$T^{\lambda}{}_{\mu\nu\eta} = R^{\lambda}{}_{\mu\nu\eta} \tag{5.54}$$

in all coordinate systems that can be connected by transformations for which the partial derivative coefficients exist. Therefore, any tensor linear in the second derivative of the metric tensor is equivalent to the Riemann curvature tensor.

Whew!

5.5 Discussion Questions and Exercises

Discussion Questions

Q5.1 In the nineteenth century, the Great Plains of North America were surveyed into 1 square mile units called "sections" (640 acres). The sections were rectangular, with survey lines running north-south and east-west. As the surveyors worked their way north, what difficulties did they encounter in trying to mark out sections of equal area? What did they do about their problem?

Q5.2 Distinguish geometrical from topological differences between an infinite plane and an infinitely long cylinder; between a solid hemisphere and a sphere; between a donut and a coffee cup.

Exercises

5.1 Show that the nonzero Ricci tensor components for the metric of Ex. 4.4 are

$$R_{00} = AN + \frac{1}{rD}(2A\dot{C} - \dot{A}C), \tag{5.55}$$

$$R_{01} = CN - \frac{A\dot{B}}{rD}, \tag{5.56}$$

$$R_{11} = -BN - \frac{1}{rD}(\dot{B}C + D'), \tag{5.57}$$

$$R_{22} = \frac{A}{D} - 1 + \frac{r}{D}(A' - \dot{C}) + \frac{r}{2D^2}(C\dot{D} - AD'), \tag{5.58}$$

$$R_{33} = R_{22}\sin^2\theta, \tag{5.59}$$

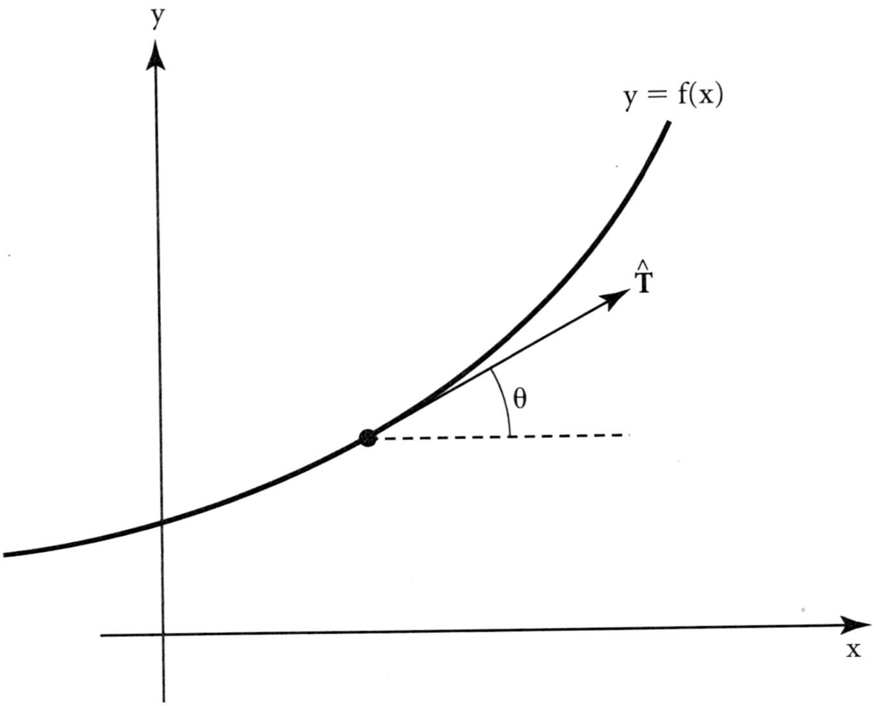

Figure 5.2: *The curve* $y = f(x)$ *and its tangent vector* $\hat{\mathbf{T}}$.

where $D = AB + C^2$, the primes denote derivatives with respect to r and overdots derivatives with respect to t, and

$$\begin{aligned} N \equiv & -\frac{A'}{rD} + \frac{1}{2D}(-A'' + 2\dot{C}' + \ddot{B}) + \frac{1}{4CD}(\dot{A}B' - A'\dot{B}) \\ & + \frac{1}{4D^2}\left[\frac{B}{C}(A'\dot{D} - \dot{A}D') + D'A' - \dot{D}\dot{B} - 2\dot{C}D'\right]. \end{aligned}$$

5.2 Calculate the Ricci tensor components $R_{\mu\nu}$ and the curvature scalar R for the metric of Ex. 4.3.

5.3 This problem invites you to work through an analogy to the Riemann tensor, which applies to any curve in a Cartesian plane. Consider a curve $y = f(x)$ in a plane mapped with (x, y) coordinates or with polar coordinates (r, θ) (see Fig. 5.2). Let $\hat{\mathbf{T}}$ be a unit vector tangent to the curve,

$$\hat{\mathbf{T}} = (\cos\theta)\hat{\mathbf{i}} + (\sin\theta)\hat{\mathbf{j}}, \tag{5.60}$$

where $\tan\theta = dy/dx \equiv y'$. Parameterize $\hat{\mathbf{T}} = \hat{\mathbf{T}}(s)$, where s is a scalar parameter, such as arc length along the curve, for which

$$ds = \sqrt{dx^2 + dy^2}. \tag{5.61}$$

(a) Evaluate a "curvature vector" $\boldsymbol{\kappa}$ defined according to

$$\begin{aligned}\boldsymbol{\kappa} &\equiv \frac{d\hat{\mathbf{T}}}{ds} \\ &= \frac{d\hat{\mathbf{T}}}{d\theta}\frac{d\theta}{ds}.\end{aligned}$$

Note that $\boldsymbol{\kappa}$ plays a role here analogous to that of the Riemann curvature tensor in higher-dimensioned spaces. Show that $\boldsymbol{\kappa}$ may be written

$$\boldsymbol{\kappa} = \frac{[-(\sin\theta)\hat{\mathbf{i}} + (\cos\theta)\hat{\mathbf{j}}]y''}{(1+y'^2)^{3/2}}, \tag{5.62}$$

where $y'' \equiv dy'/dx$.
(b) We can also introduce a "curvature scalar" $\kappa \equiv |\boldsymbol{\kappa}|$. The "radius of curvature" ρ is its inverse, $\rho = 1/\kappa$. If $s = s(t)$, where t denotes time, show that

$$\kappa = \frac{\dot{\theta}}{\dot{s}}, \tag{5.63}$$

where $\dot{\theta} \equiv d\theta/dt$, and that

$$\kappa = \frac{|\ddot{x}y - \ddot{y}x|}{(\dot{x}^2 + \dot{y}^2)^{3/2}}, \tag{5.64}$$

which is an analogue of the Riemann curvature scalar R.
(c) Verify that the evaluation of $\boldsymbol{\kappa}, \kappa$, and ρ gives the expected results for a circle of radius a centered on the origin.
(d) Calculate $\boldsymbol{\kappa}, \kappa$, and ρ at the point $(1, 1)$ on the curve $y = x^2$.
For more on the curvature vector and scalar in the Euclidean plane, see Leithold (Sec. 15.9) or Wardle (Ch. 1); for the Riemann curvature tensor on surfaces, see Weinberg (Sec. 1.1 and p. 144), Laugwitz (Sec. 11.6), or Synge and Schild (Sec. 3.4).

5.4 Demonstrate these symmetry properties of the Riemann curvature tensor:
(a) $R_{\alpha\beta\mu\nu} = -R_{\beta\alpha\mu\nu}$;
(b) $R_{\alpha\beta\mu\nu} = -R_{\alpha\beta\nu\mu}$;
(c) $R_{\alpha\beta\mu\nu} = R_{\mu\nu\alpha\beta}$.

5.5 Compute $R_{\mu\nu}$ and R on a spherical shell of radius a.

5.6 Write at least some of the components of $R^{\mu}{}_{\nu\rho\sigma}$ in plane polar coordinates.

5.7 In 1917, Albert Einstein solved various "problems at infinity" (such as Olber's paradox) which existed in a three-dimensional, eternal, static, infinite universe assumed by Newtonian cosmology: Einstein abolished infinity! He hypothesized that cosmological three-dimensional space is actually the surface of a sphere of radius a that exists in a four-dimensional Euclidean space. Thus, in the spatial part of spacetime, any point in the universe that we would label with coordinates (x, y, z) would actually reside on the surface of this hypersphere, which from the perspective of the four-dimensional embedding space would be described by the equation

$$a^2 = x^2 + y^2 + z^2 + w^2, \tag{5.65}$$

which requires the introduction of a new coordinate axis w.
(a) Show that the metric of the space on the hypersphere (and thus the geometry of our universe in Einstein's model) may be written, in terms of three-dimensional spherical coordinates $(x, y, z) \to (r, \theta, \phi)$, as

$$ds^2 = \left(1 - \frac{a^2}{r^2}\right)^{-1} dr^2 + r^2 d\theta^2 + r^2 \sin^2\theta d\phi^2. \tag{5.66}$$

(b) Let some point be chosen to be the origin of three-dimensional space. Circumnavigate the origin on a shell of fixed radius to measure the circumference C, and then define the value of the shell's r-coordinate according to $r \equiv C/2\pi$. If a traveler moves from a shell with coordinate r_1 to another shell with coordinate r_2, what is the difference between $r_2 - r_1$ and the distance between the two shells as measured directly with a meterstick held radially?
(c) Let $w = a\cos\chi$ and define an angle χ where $0 \le \chi \le \pi$. Show that the metric for four-dimensional space can be written

$$ds^2 = a^2 d\chi^2 + a^2 \sin^2\chi d\theta^2 + a^2 \sin^2\chi \sin^2\theta d\phi^2. \tag{5.67}$$

(d) Show that the area of the hypersphere's spherical surface equals $2\pi^2 a^3$, which is the volume of the three-dimensional universe in this model.

5.8 Two particles fall freely through spacetime and therefore follow geodesics. Suppose they are falling radially toward the center of the Earth. An observer riding along with either particle would notice them gradually approaching one another (even neglecting the particle's mutual gravitation), since the radial free-fall lines converge toward Earth's center. Let $\zeta^\mu = x_2^\mu - x_1^\mu$ be the component of a 4-vector from particle 1 to nearby particle 2. For instance, an astronaut in free fall toward the Earth could release two pennies inside the spacecraft and closely measure the distance between them. Show to first order in ζ^μ that the line connecting the two pennies respects the "equation of geodesic deviation,"

$$\frac{D^2\zeta^\lambda}{D\tau^2} + R^\lambda{}_{\mu\nu\rho}\frac{dx^\mu}{d\tau}\zeta^\nu\frac{dx^\rho}{d\tau} = 0 \tag{5.68}$$

(see Weinberg, p. 148; or Misner, Thorne, and Wheeler, p. 34).

5.9 (a) Show that the Ricci tensor may be written

$$R_{\mu\nu} = \partial_\nu \Gamma^\rho{}_{\mu\rho} - \partial_\rho \Gamma^\rho{}_{\mu\nu} + \Gamma^\lambda{}_{\mu\rho}\Gamma^\rho{}_{\lambda\nu} - \Gamma^\lambda{}_{\mu\nu}\Gamma^\rho{}_{\lambda\rho}. \tag{5.69}$$

(b) Derive a generic expression for the curvature scalar $R = R^\mu{}_\mu$ in terms of the metric tensor and affine connection coefficients.

Chapter 6

Covariance Applications

In this chapter we demonstrate how important principles of physics, when expressed in as tensor equations, can lay important insights in one's path.

6.1 Covariant Electrodynamics

It is well known that Maxwell's electrodynamics–as usually understood at present–when applied to moving bodies, leads to asymmetries that do not seem to be inherent in the phenomena"
–Albert Einstein's introduction to "On the Electrodynamics of Moving Bodies" (1905)

As claimed in Section 1.8, an electric field $\mathbf{E}$ and a magnetic field $\mathbf{B}$, measured in the Lab Frame, can be transformed into the fields $\mathbf{E}'$ and $\mathbf{B}'$ measured in the Coasting Rocket Frame. Now we have the tools to carry out the transformation efficiently. All we have to do is write the fields as tensors in the Lab Frame, and then apply a Lorentz transformation that carries them into the Coasting Rocket Frame.

Maxwell's equations are four coupled first-order partial differential equations for $\mathbf{E}$ and $\mathbf{B}$. The two inhomogeneous Maxwell equations, which relate the fields to the charged particles that produce them, include Gauss's Law for the electric field $\mathbf{E}$,

$$\nabla \cdot \mathbf{E} = \frac{\rho}{\epsilon_o}, \tag{6.1}$$

and the Ampère-Maxwell law,

$$\nabla \times \mathbf{B} = \mu_o \mathbf{j} + \mu_o \epsilon_o \frac{\partial \mathbf{E}}{\partial t}. \tag{6.2}$$

The homogeneous Maxwell equations include Gauss's law for the magnetic field,

$$\nabla \cdot \mathbf{B} = 0, \tag{6.3}$$

and the Faraday-Lenz law,

$$\nabla \times \mathbf{E} = -\frac{\partial \mathbf{B}}{\partial t}. \tag{6.4}$$

Although **E** and **B** are vectors *in space*, they are not vectors *in spacetime*, so expressions written in terms of them will not be relativistically covariant. The avenue through the electromagnetic potentials offers a straightforward path to general covariance, as we shall see. First, we need the potentials themselves.

Since the divergence of a curl vanishes identically, Gauss's law for **B** implies that the magnetic field is the curl of a vector potential **A**,

$$\mathbf{B} = \nabla \times \mathbf{A}. \tag{6.5}$$

Put this into the Faraday-Lenz law, which becomes

$$\nabla \times \left[\mathbf{E} + \frac{\partial \mathbf{A}}{\partial t}\right] = 0. \tag{6.6}$$

Because the curl of a gradient vanishes identically, this implies that a potential ϕ exists such that

$$\mathbf{E} = -\nabla\phi - \frac{\partial \mathbf{A}}{\partial t}. \tag{6.7}$$

The fields **A** and ϕ are the electromagnetic potentials. When **E** and **B** are written in terms of these potentials, the inhomogeneous Maxwell equations become a pair of coupled second-order partial differential equations. In particular, upon substituting for **E** and **B** in terms of the potentials, the Ampère-Maxwell law becomes

$$\nabla^2\mathbf{A} - \frac{1}{c^2}\frac{\partial^2 \mathbf{A}}{\partial t^2} = -\mu_o\mathbf{j} + \frac{1}{c^2}\nabla\left(\frac{\partial \phi}{\partial t}\right) + \nabla(\nabla \cdot \mathbf{A}), \tag{6.8}$$

and Gauss's law for **E** becomes

$$\nabla^2\phi = -\frac{\rho}{\epsilon_o} - \frac{\partial}{\partial t}(\nabla \cdot \mathbf{A}). \tag{6.9}$$

So far **A** and ϕ are defined terms of their derivatives. This means that **A** and ϕ are ambiguous because one could always add to them a term that vanishes when taking the derivatives. Changing that additive term is called a "gauge transformation" on the potentials. For a gauge transformation to work for both **A** and ϕ simultaneously, it takes the form, for an arbitrary function χ,

$$\mathbf{A}' = \mathbf{A} + \nabla\chi \tag{6.10}$$

and

$$\phi' = \phi - \frac{\partial \chi}{\partial t}. \tag{6.11}$$

The **E** and **B** fields are unchanged because the arbitrary function χ drops out, leaving $\mathbf{E}' = \mathbf{E}$ and $\mathbf{B}' = \mathbf{B}$, as you can readily verify.

We can use this freedom to choose the divergence of **A** to be whatever we want. For instance, if you are given an **A** for which $\nabla \cdot \mathbf{A} = \alpha$, and if you prefer the divergence of **A** to be something else, say, β, then you merely make a gauge transformation to a new vector potential $\mathbf{A}' = \mathbf{A} + \nabla\chi$ and require $\nabla \cdot \mathbf{A}' = \beta$, which in turn requires that χ satisfies

$$\nabla^2\chi = \alpha - \beta. \tag{6.12}$$

To get the divergence of the vector potential that you want, merely solve this Poisson's equation for the necessary gauge-shifting function χ. Since the χ is no longer arbitrary, one speaks of "choosing the gauge."

The differential equations for **A** and ϕ neatly uncouple if we choose the "Lorentz gauge,"

$$\nabla \cdot \mathbf{A} = -\frac{1}{c^2}\frac{\partial\phi}{\partial t}. \tag{6.13}$$

This so-called Lorentz condition turns the coupled differential equations into uncoupled inhomogeneous wave equations for **A** and ϕ:

$$\begin{aligned}
\nabla^2\mathbf{A} - \frac{1}{c^2}\frac{\partial^2\mathbf{A}}{\partial t^2} &= -\mu_c\mathbf{j} \\
\nabla^2\phi - \frac{1}{c^2}\frac{\partial^2\phi}{\partial t^2} &= -\frac{\rho}{\epsilon_c}.
\end{aligned}$$

Can we now write electrodynamics covariantly in the spacetime of special relativity? Let us absorb c into the definition of the time coordinate. The potentials **A** and ϕ may be gathered into a 4-vector in spacetime,

$$A^\mu = (A^0, A^1, A^2, A^3) = (\phi, \mathbf{A}), \tag{6.14}$$

as may the 4-vector of charge density and current density (with $c = 1$),

$$j^\mu = \rho u^\mu = (j^0, j^1, j^2, j^3) = (\rho, \mathbf{j}). \tag{6.15}$$

How do we know that the A^μ and the j^μ are components of first-order tensors under Lorentz transformations? Of course, it is by how they transform; see Ex. 6.8.

In Minkowskian spacetime, the duals to the 4-vectors are

$$\begin{aligned}
A_\mu &= (\phi, -\mathbf{A}) \\
j_\mu &= (\rho, -\mathbf{j}).
\end{aligned}$$

Let us also recall the two forms of the gradient,

$$\begin{aligned}
\partial_\mu &= \left(\frac{\partial}{\partial t}, \nabla\right) \\
\partial^\mu &= \left(\frac{\partial}{\partial t}, -\nabla\right).
\end{aligned}$$

Because electric charge is locally conserved, it respects the equation of continuity

$$\nabla \cdot \mathbf{j} + \frac{\partial \rho}{\partial t} = 0, \tag{6.16}$$

which may be written covariantly,

$$\partial_\mu j^\mu = 0. \tag{6.17}$$

The gauge transformation may be written with spacetime covariance as

$$A'^\mu = A^\mu - \partial^\mu \chi \tag{6.18}$$

and the Lorentz condition expressed as

$$\partial_\mu A^\mu = 0. \tag{6.19}$$

Both inhomogeneous wave equations for $\mathbf{A}$ and ϕ are now subsumed into

$$\partial_\mu \partial^\mu A^\nu = \mu_o j^\nu, \tag{6.20}$$

where we have used

$$\mu_o \epsilon_o = \frac{1}{c^2} \tag{6.21}$$

so that, with $c = 1$, we may write $1/\epsilon_o = \mu_o$.

However, we are still not satisfied: although this wave equation is covariant with respect to Lorentz transformations among inertial reference frames, it is *not* covariant with respect to a gauge transformation within a given spacetime reference frame, because Eq. (6.20) holds only in the Lorentz gauge. Can we write for the electromagnetic field a wave equation that holds *whatever* the gauge, and thus is gauge-covariant as well as Lorentz-covariant? To pursue this line of reasoning, return to the coupled equations for $\mathbf{A}$ and ϕ as they appeared before a gauge was chosen. Using the upper- and lower-index forms of the gradient, and separating the space ($\mu = k = 1, 2, 3$) and time ($\mu = 0$) coordinates explicitly, Eq. (6.9) says that

$$\partial^k \partial_k A^0 - \partial^0 (\partial_k A^k) = \mu_o j^0, \tag{6.22}$$

which can be regrouped into

$$\partial_k (\partial^k A^0 - \partial^0 A^k) = \mu_o j^0. \tag{6.23}$$

The equation for the $n = 1, 2, 3$ components of Eq. (6.8) becomes

$$\partial_k \partial^k A^n + \partial_0 \partial^0 A^n = \mu_o j^n + \partial^n (\partial_\mu A^\mu), \tag{6.24}$$

or, upon rearranging,

$$\partial_\mu (\partial^\mu A^n - \partial^n A^\mu) = \mu_o j^n. \tag{6.25}$$

Eqs. (6.23) and (6.25) suggest introducing a second-order tensor, the "Faraday tensor" defined by a covariant curl,

$$F^{\mu\nu} \equiv \partial^\mu A^\nu - \partial^\nu A^\mu, \tag{6.26}$$

which is antisymmetric:

$$F^{\mu\nu} = -F^{\nu\mu}. \tag{6.27}$$

Comparing $F^{\mu\nu}$ to $\mathbf{E} = -\nabla\phi - \frac{\partial \mathbf{A}}{\partial t}$ and $\mathbf{B} = \nabla \times \mathbf{A}$, the components of the Faraday tensor can be written in terms of the ordinary xyz components of $\mathbf{E}$ and $\mathbf{B}$ (where the xyz subscripts are merely labels, not covariant vector indices):

$$F^{\mu\nu} = \begin{pmatrix} 0 & -E_x & -E_y & -E_z \\ E_x & 0 & -B_z & B_y \\ E_y & B_z & 0 & -B_x \\ E_z & -B_y & B_x & 0 \end{pmatrix}. \tag{6.28}$$

The Faraday tensor is invariant under a gauge transformation *and* transforms as a second-order tensor under spacetime coordinate transformations, as you can easily verify. In terms of the Faraday tensor, the inhomogeneous Maxwell equations can be written as one tensor equation with four components,

$$\partial_\mu F^{\mu\nu} = \mu_o j^\nu, \tag{6.29}$$

where $\nu = 0, 1, 2, 3$. The homogeneous equations can also be written in terms of the Faraday tensor (note the cyclic appearance of the indices),

$$\partial_\mu F_{\nu\lambda} + \partial_\nu F_{\lambda\mu} + \partial_\lambda F_{\mu\nu} = 0, \tag{6.30}$$

where

$$F_{\mu\nu} = g_{\mu\rho} g_{\nu\sigma} F^{\rho\sigma}. \tag{6.31}$$

These expressions are not only covariant with respect to spacetime transformations but also gauge invariant.

Now at last the transformation of $\mathbf{E}$ and $\mathbf{B}$ from the Lab Frame to the Coasting Rocket Frame proceeds straightforwardly: The new Faraday tensor is made from components of the old ones according to the usual spacetime transformation rule for second-order tensors,

$$F'^{\mu\nu} = \frac{\partial x'^\mu}{\partial x^\rho} \frac{\partial x'^\nu}{\partial x^\sigma} F^{\rho\sigma}. \tag{6.32}$$

Given the original electric and magnetic fields and the transformation matrix elements $\frac{\partial x'^\mu}{\partial x^\rho}$, the new electric and magnetic fields follow at once.

For connecting inertial reference frames, the Λ-matrices are elements of the Poincaré group, which consists of Lorentz boosts, rotations of axes, and spacetime origin translations. We saw that for the simple Lorentz boost with x and x' axes parallel, and no origin offset,

$$\begin{aligned} t' &= \gamma_r(t - v_r x) \\ x' &= \gamma_r(x - v_r t) \\ y' &= y \\ z' &= z, \end{aligned}$$

expressed in the transformation matrix

$$\Lambda = \begin{pmatrix} \gamma_r & -\gamma_r v_r & 0 & 0 \\ -\gamma_r v_r & \gamma_r & 0 & 0 \\ 0 & 0 & 1 & 0 \\ 0 & 0 & 0 & 1 \end{pmatrix}, \tag{6.33}$$

where $\frac{\partial x'^{\rho}}{\partial x^{\sigma}}$ sits in the matrix element entry at the intersection of row ρ and column σ.

With this simple Lorentz transformation used in Eq. (6.32), we find for the new electric field

$$E'_x = E_x \tag{6.34}$$

$$E'_y = \gamma_r(E_y - v_r B_z) \tag{6.35}$$

$$E'_z = \gamma_r(E_z + v_r B_y) \tag{6.36}$$

and for the new magnetic field

$$B'_x = B_x \tag{6.37}$$

$$B'_z = \gamma_r(E_z + v_r B_y) \tag{6.38}$$

$$B'_z = \gamma_r(B_z - vE_y). \tag{6.39}$$

Recall how the rotation matrix about one axis could be generalized to an arbitrary rotation by using Euler angles. Similarly, for an arbitrary constant velocity $\mathbf{v}_r$ that boosts the Coasting Rocket Frame relative to the Lab Frame, the boost matrix works out to be the following (where, to reduce notational clutter, the r subscripts have been dropped from γ_r and the $\mathbf{v}_r$ components in the matrix elements):

$$\Lambda(\mathbf{v}_r) = \begin{pmatrix} \gamma & -\gamma v_x & -\gamma v_y & -\gamma v_z \\ -\gamma v_x & 1 + \frac{(\gamma-1)v_x^2}{v^2} & \frac{(\gamma-1)v_x v_y}{v^2} & \frac{(\gamma-1)v_x v_z}{v^2} \\ -\gamma v_y & \frac{(\gamma-1)v_x v_y}{v^2} & 1 + \frac{(\gamma-1)v_y^2}{v^2} & \frac{(\gamma-1)v_y v_z}{v^2} \\ -\gamma v_z & \frac{(\gamma-1)v_x v_z}{v^2} & \frac{(\gamma-1)v_y v_z}{v^2} & 1 + \frac{(\gamma-1)v_z^2}{v^2} \end{pmatrix}, \tag{6.40}$$

where

$$\gamma = \frac{1}{\sqrt{1 - v^2}}. \tag{6.41}$$

For this Lorentz boost, Eq. (6.32) gives the result claimed in Section 1.8 (here with $c = 1$),

$$\mathbf{E}' = \gamma\left[\mathbf{E} + (\mathbf{v} \times \mathbf{B})\right] - \frac{\gamma^2}{\gamma+1}\mathbf{v}(\mathbf{v} \cdot \mathbf{E}),$$

$$\mathbf{B}' = \gamma\left[\mathbf{B} - (\mathbf{v} \times \mathbf{E})\right] - \frac{\gamma^2}{\gamma+1}\mathbf{v}(\mathbf{v} \cdot \mathbf{B}).$$

6.2 General Covariance and Gravitation

The principle of equivalence demands that in dealing with Galilean regions we may equally well make use of non-inertial systems, that is, such co-ordinate systems as, relatively to inertial systems, are not free from acceleration and rotation. If, further, we are going to do away completely with the vexing question as to the objective reason for the preference of certain systems of co-ordinates, then we must allow the use of arbitrarily moving systems of co-ordinates.

–Albert Einstein, *The Meaning of Relativity*

Newtonian gravitation theory follows from two empirically based principles, Newton's law of universal gravitation and the principle of superposition. In terms of the Newtonian gravitational field $\mathbf{g}$, its divergence and curl field equations are

$$\nabla \cdot \mathbf{g} = -4\pi G\rho \tag{6.42}$$

and

$$\nabla \times \mathbf{g} = 0, \tag{6.43}$$

where G denotes Newton's gravitational constant and ρ the mass density. From the vanishing of the curl we may infer that a gravitational potential ϕ exists for which $\mathbf{g} = -\nabla\phi$, which in the divergence equation yields Poisson's equation to be solved for the static gravitational potential,

$$\nabla^2\phi = 4\pi G\rho. \tag{6.44}$$

The principle of the equivalence of gravitational and inertial mass says that gravitation can be locally transformed away by going into a local free-fall frame. Thus, gravitation can be "turned off and on" by changing the reference frame, which sounds like a job for tensors! This offers motivation enough for finding a generally relativistic theory of gravitation. Further motivation comes from the realization that Newtonian gravitation boldly assumes instantanous action-at-a-distance, contrary to the special theory of relativity, which asserts that no signal may travel faster than the speed of light in vacuum.

Special relativity and Newtonian relativity share a preference for inertial reference frames. Special relativity made its appearance in 1905 when Albert Einstein published "On the Electrodynamics of Moving Bodies." During the next decade, he worked through monumental difficulties to generalize the program to accelerated reference frames–when incidentally he had to master tensor calculus. Besides generalizing special relativity to general relativity, thanks to the principle of equivalence it also generalized Newtonian gravitation to a generally relativistic theory of gravity, by carrying out a coordinate transformation! The transformation takes one from a locally inertial reference frame where all particles fall freely, with no discernible gravity, to any other reference frame accelerated relative to the inertial one. Due to the equivalence principle, the acceleration is locally equivalent to gravity.

Figure 6.1: *Georg Friedrich Bernhard Riemann (1826-1866). Photo from Wikimedia Commons.*

The necessary tools grew out of the non-Euclidean geometries that had been developed in preceding decades by Nikoli Lobachevski (1792-1856), Janos Bolyai (1802-1860), Karl Friedrich Gauss (1777-1855), and culminating in the work of Georg Friedrich Bernhard Riemann (1826-1866) (see Fig. 6.1). Riemann generalized and expanded on the work of his predecessors, extending the study of geometry on curved surfaces to geometry in spaces of n dimensions and variable curvature. The language of tensors offered the ideal tool for Einstein to use in grappling with his problem. Through this language, with its Riemann curvature tensor, Einstein's theory leads to the notion of gravitation as the "curvature of spacetime."

For such a notion to be made quantitative, so that its inferences could be tested against the real world, the working equations of Newtonian gravitational dynamics had to be extended, to be made generally covariant. Start in an inertial freely falling Minkowskian frame that uses coordinates X^μ and coordinate velocities U^μ, so that for a freely floating body,

$$\frac{dU^\mu}{d\tau} = 0. \tag{6.45}$$

When the covariant derivative replaces the ordinary derivative, and the local coordinates X^μ (with their metric tensor $\eta_{\mu\nu}$) are replaced with curvilinear coordinates x^μ (with their metric tensor $g_{\mu\nu}$), these changes turn $dU^\mu/d\tau = 0$ into

$$\frac{Du^\mu}{D\tau} = 0. \tag{6.46}$$

Writing out the full expression for the covariant derivative, this becomes

$$\frac{du^\mu}{d\tau} + \Gamma^\mu{}_{\rho\sigma} u^\rho u^\sigma = 0, \tag{6.47}$$

which we have seen before, e.g., Eq. (4.18).

A similar program needs to be done with the Newtonian gravitational field, applying the principle of general covariance to gravitation itself. Since Poisson's equation is the differential equation of Newtonian gravitation, following Einstein, we will extend it, so the end result will be a generally covariant tensor equation. Recall from Section 4.3 that g_{00}, in the Newtonian limit here denoted $\tilde{g}_{00}$, was shown to be approximately

$$\tilde{g}_{00} \approx 1 + 2\phi, \tag{6.48}$$

in terms of which Poisson's equation, Eq. (6.44), takes the form

$$\nabla^2 \tilde{g}_{00} = 8\pi G\rho. \tag{6.49}$$

I put the tilde on $\tilde{g}_{00}$ here to remind us that this is merely the Newtonian limit of the time-time component of the exact g_{00}. Clearly the way to start extending this to a generally relativistic equation is to see in the $\tilde{g}_{00}$ of Eq. (6.49) a special case of $g_{\mu\nu}$. We also have to construct generally covariant replacements in spacetime for the Laplacian and for the mass density.

For replacing the mass density we take a hint from special relativity, which shows mass to be a form of energy. Thus, the extension of the mass density ρ will be energy density or other dynamical variables with the same dimensions, such as pressure. The stress tensors $T_{\mu\nu}$ of fluid mechanics and electrodynamics, which describe energy and momentum transport of a fluid or electromagnetic fields, are second-order tensors. Their diagonal elements are typically energy density or pressure terms (e.g., recall Section 2.4 and Ex. 2.14). Therefore, the ρ in Poisson's equation gets replaced by a stress tensor $T_{\mu\nu}$, the particulars of which depend on the system whose energy and momentum serve as the sources of gravitation. For instance, special relativity shows that for an ideal continuous relativistic fluid, the stress tensor is

$$T_{\mu\nu} = \eta_{\mu\nu} P + (\rho + P) \frac{dX^\mu}{d\tau} \frac{dX^\nu}{d\tau}, \tag{6.50}$$

where $\eta_{\mu\nu}$ denotes the metric tensor in flat Minkowskian spacetime, with P and ρ the pressure and energy density, respectively (replace P with P/c^2 when using conventional units). To generalize this $T_{\mu\nu}$ to a system of general curvilinear coordinates x^μ with metric tensor $g_{\mu\nu}$, we write, by general covariance,

$$T_{\mu\nu} = g_{\mu\nu} P + (\rho + P) \frac{Dx^\mu}{D\tau} \frac{Dx^\nu}{D\tau}. \tag{6.51}$$

Various equations of state exist that relate P and ρ. Some of them take the form (in conventional units)

$$P = w\rho c^2, \tag{6.52}$$

where w is a dimensionless constant. For instance, with static dust $w = 0$, for electromagnetic radiation $w = \frac{1}{3}$, and for the enigmatic "dark energy" of cosmology $w = -1$.

Since energy and momentum are locally conserved, an equation of continuity holds for the stress tensor, which requires in general the covariant derivative,

$$D^\mu T_{\mu\nu} = 0. \tag{6.53}$$

As noted above, on the left-hand side of the extended Poisson's equation $\tilde{g}_{00}$ will get replaced by $g_{\mu\nu}$, and the second derivatives of the Laplacian must be extended to second covariant derivatives with respect to spacetime coordinates. If the right side of the covariant extension of Poisson's equation is going to be a second-order tensor, then whatever replaces the Laplacian on the left-hand side must also be a tensor of second order. To make his theory no more complicated than necesary, Einstein hypothesized that the differential operator should be linear in the second derivative of $g_{\mu\nu}$, analogous to how the Laplacian is linear in its second derivatives. Since the Riemann tensor is the only tensor linear in the second covariant derivative, Einstein turned to it. Requiring its second-order version, the Ricci tensor $R_{\mu\nu}$ would appear. So could its completely contracted version, the curvature scalar R, provided that it appears with the metric tensor as the combination $Rg_{\mu\nu}$ to offer another order-2 tensor based on the Riemann tensor. In addition, general covariance also allows term $\Lambda g_{\mu\nu}$, where Λ is a constant unrelated to the Riemann curvature tensor.

With these considerations, Einstein had a template for a generally covariant extension of Poisson's equation,

$$AR_{\mu\nu} + Bg_{\mu\nu}R + \Lambda g_{\mu\nu} = 8\pi G T_{\mu\nu}, \tag{6.54}$$

where A, B, and Λ are constants. It remained to impose various constraints to determine these constants.

The constant Λ carries an interesting history. Einstein originally set it equal to zero, since he was committed to the notion of gravitation as the curvature of spacetime, and curvature is carried by $R_{\mu\nu}$ and R. The initial applications that Einstein made with his equations included explaining the long-standing problem of the anomalous precession of Mercury's orbit, and a new prediction of the amount of deflection of light rays by the Sun, which was affirmed in 1919. These results were astonishingly successful with $\Lambda = 0$. When Einstein applied the general theory to cosmology in 1917, he assumed a static universe, which the precision of contemporary astronomical data seemed to justify. But his cosmological equations were not consistent unless he brought a nonzero Λ back into the program. Thus, Λ became known as the "cosmological constant." It is negligible on the scale of a solar system or a galaxy, but could be appreciable on the scale of the entire cosmos. Significantly, it enriched the discussion of a static versus an expanding universe, and it has re-emerged as a question of capital importance in cosmology today. But pursuing this fascinating topic would take us too far afield. Our task is to see how Einstein used tensor equations with their general covariance to produce the field equations of general relativity in the first place. Therefore, as Einstein did originally, let us set $\Lambda = 0$. The $\Lambda g_{\mu\nu}$ term can always be added back in afterward, because the theoretical constraints that determine A and B say nothing about Λ anyway. It remains a parameter to be fit to data, or assumed in a model.

Without Λ, the template for the relativistic equations of gravitation is

$$AR_{\mu\nu} + Bg_{\mu\nu}R = 8\pi G T_{\mu\nu}. \tag{6.55}$$

This is a set of 16 equations, but only 10 are independent because the Ricci, metric, and stress tensors are symmetric.

The conditions to determine A and B include
(1) local conservation of energy and momentum, expressed by the vanishing of the covariant divergence of the stress tensor, $D^{\mu}T_{\mu\nu} = 0$;
(2) the Bianchi identities for the Riemann tensor; and
(3) the Newtonian limit.

Starting with the divergence, setting $T^{\mu\nu}{}_{;\mu} = 0$ requires

$$AR^{\mu\nu}{}_{;\mu} + Bg^{\mu\nu}{}_{;\mu}R + Bg^{\mu\nu}R_{;\mu} = 0. \tag{6.56}$$

But $g^{\mu\nu}{}_{;\nu} = 0$ (recall Ex. 4.2). In addition, the Bianchi identity

$$R^{\mu\nu}{}_{;\nu} = \frac{1}{2}g^{\mu\nu}R_{;\nu} \tag{6.57}$$

yields $B = -A/2$. Thus, the field equations so far read

$$A\left(R_{\mu\nu} - \frac{1}{2}g_{\mu\nu}R\right) = 8\pi G T_{\mu\nu}. \tag{6.58}$$

The Newtonian nonrelativisitic, static, weak-field limit requires $A = -1$ (see Ex. 6.3), leaving us with Einstein's field equations

$$R_{\mu\nu} - \frac{1}{2}g_{\mu\nu}R = -8\pi G T_{\mu\nu}. \tag{6.59}$$

In deriving his field equations for gravitation, Einstein applied a procedure that offers an algorithm for doing physics in the presence of gravitation, if one first knows how to do physics in the absence of gravitation. The principle of the equivalence of gravitational and inertial mass assures us that we can always find a reference frame for doing physics in the absence of measurable gravitation: for a sufficiently short interval of time, and over a sufficiently small region of space (the latter to make tidal forces negligible), we can go into a free-fall frame. For instance, if my chair were to somehow instantaneously dissolve, then for a brief moment, over a small region, I would be in a state of free-fall! Suppose in this locally inertial frame we have some important physics equation, perhaps some quantity J is related to the derivative of some observable F. We can represent the overall idea with a cartoon equation (suppressing any indices)

$$\partial F(X) = J(X), \tag{6.60}$$

where "∂" stands for whatever derivatives are necessary. For instance, ∂ might be the Laplacian, F an electromagnetic potential, and J an electric current density, so that $\partial F = J$ stands for an equation from electrodynamics. Furthermore, in this locally inertial frame, the coordinates are X^μ with metric tensor $\eta_{\mu\nu}$. For the next step of the algorithm to work, it is crucial that $\partial F = J$ be an equality between *tensors*. For then we "turn on gravitation" by transforming into a general coordinate system with coordinates x^μ and metric tensor $g_{\mu\nu}$ and replace the usual derivative ∂ with the appropriate covariant derivative D, to write

$$DF(x) = J(x). \tag{6.61}$$

Now we are doing physics in the presence of gravitation, because the D contains the affine connection terms (and thus gravitation) within it, because $D \sim \partial + \Gamma - term$.

6.3 Discussion Questions and Exercises

Discussion Questions

Q6.1 Find out whatever you can about "minimal coupling" as used in electrodynamics, where a particle of mass m carrying electric charge q moves in an electromagnetic field described by the vector potential $\mathbf{A}$. The minimal coupling algorithm says to replace (in the nonrelativistic case) the particle's momentum $m\mathbf{v}$ with $m\mathbf{v} + q\mathbf{A}$. How is this procedure analogous to the algorithm used in the principle of general covariance?

Q6.2 Suppose you write a particle's equation of motion in an inertial reference frame, within the paradigms of Newtonian relativity. Let coordinates x^k

be rectangular Cartesian coordinates, with time intervals taken to be invariant. Discuss the validity of the following procedure: To jump from the inertial frame to a rotating reference frame, where the rotating frame spins with constant angular velocity $\boldsymbol{\omega}$ relative to the inertial frame, and the x'^k are the coordinates within the rotating frame, start with the equations of motion as written in the inertial frame. Then change $\mathbf{r}$ to $\mathbf{r}'$, and replace the velocity $\mathbf{v}$ with $\mathbf{v}' + (\boldsymbol{\omega} \times \mathbf{r}')$. Does this give the correct equation of motion *within the rotating system of coordinates*? What are the so-called fictitious forces that observers experience in the rotating frame? This program is most efficiently done with Lagrangian formalism, where the equation of motion follows from the Euler-Lagrange equation,

$$\frac{\partial L}{\partial x^\mu} = \frac{d}{dt}\left(\frac{\partial L}{\partial \dot{x}^\mu}\right), \tag{6.62}$$

with $\dot{x}^\mu \equiv dx^\mu/dt$ and L the Lagrangian, in this context the difference between the kinetic and potential energy,

$$L(x^\mu, \dot{x}^\mu, t) = \frac{1}{2}m\mathbf{v}\cdot\mathbf{v} - U(\mathbf{r}) \tag{6.63}$$

(see Dallen and Neuenschwander).

Q6.3 Discuss whether Lagrangian and Hamiltonian dynamics are already "generally covariant." Can covariance under spacetime transformations *and* gauge transformations be unified into a common formalism (see Neuenschwander)?

Q6.4 Suppose you are given a vector potential $\mathbf{A}$ and its divergence is α. Show that one can perform a gauge transformation to a different vector potential $\mathbf{A}' = \mathbf{A} + \nabla\chi$ that maintains the same divergence as $\mathbf{A}$. What equation must be satisfied by χ? To choose the divergence of $\mathbf{A}$ is to "choose a gauge," and such a choice is really a *family* of vector potentials, all with the same gauge.

Q6.5 When extending Poisson's equation into something generally covariant, why didn't Einstein merely make use of the Laplacian made of covariant derivatives, Eq. (4.92), and let it go at that?

Q6.6 In the stress tensor for an ideal relativistic fluid,

$$T_{\mu\nu} = g_{\mu\nu}P + (\rho + P)\frac{Dx^\mu}{D\tau}\frac{Dx^\nu}{D\tau}, \tag{6.64}$$

what is implied when the covariant derivatives in curvilinear coordinates are written out? Does a source of gravity, such as the relativistic fluid, depend on the local graviation? Comment on the nonlinearity of Einstein's field equations.

Exercises

6.1 Starting with the electrostatic field equations in vacuum (with no matter present other than the source charges),

$$\nabla \cdot \mathbf{E} = \frac{\rho}{\epsilon_o} \tag{6.65}$$

and

$$\nabla \times \mathbf{E} = 0, \tag{6.66}$$

write them in terms of Faraday tensor components and appropriate spatial derivatives. Then allow the static equations to become covariant in Minkowskian spacetime. Show that "electrostatics plus Lorentz covariance among inertial reference frames" yields all of Maxwell's equations (see Kobe).

6.2 Starting with the field equations of magnetostatics in vacuum,

$$\nabla \cdot \mathbf{B} = 0 \tag{6.67}$$

and

$$\nabla \times \mathbf{B} = \mu_o \mathbf{j}, \tag{6.68}$$

write them using Faraday tensor components and appropriate representations of the spatial derivatives. Then allow the static equations to become covariant in Minkowskian spacetime. Show that "magnetostatics plus Lorentz covariance among inertial reference frames" yields all of Maxwell's equations (see Turner and Neuenschwander).

6.3 Show that $A = -1$ in the Newtonian nonrelativistic, static, weak-field limit of Eq. (6.58). Do this by starting with Einstein's field equations (without Λ) with A to be determined,

$$A(R_{\mu\nu} - \frac{1}{2} g_{\mu\nu} R) = 8\pi G T_{\mu\nu}, \tag{6.69}$$

and require it to reduce to Poisson's equation for the gravitational potential.
(a) First, note that, typically, stress tensors are energy densities in their time-time components, $T_{00} = \rho$, and their time-space components are proportional to a particle's velocity. Therefore, for $v \ll 1$, we may say that

$$|T_{ij}| \ll |T_{00}|, \tag{6.70}$$

where i and j label spatial components. Show from Einstein's field equations that this implies

$$R_{ij} \approx \frac{1}{2} g_{ij} R. \tag{6.71}$$

(b) Next consider $R = R^{\mu}{}_{\mu} = g^{\mu\nu} R_{\nu\mu}$. Also recall that, in a weak field, $g^{\mu\nu} \approx \eta^{\mu\nu}$. Therefore, show that

$$R \approx R_{00} - R^{i}{}_{i}. \tag{6.72}$$

(c) From parts (a) and (b), show that

$$R = R_{00} + \frac{3}{2}R \tag{6.73}$$

and thus $R = -2R_{00}$.
(d) From the foregoing, the only component of the field equations which does not collapse to $0 = 0$ in this situation is

$$A(R_{00} - \frac{1}{2}\eta_{00}R) = 8\pi G\rho. \tag{6.74}$$

From the Ricci tensor (Ex. 5.9) and allowing only static fields, so that time derivatives vanish, show that

$$R_{00} = -\partial_\mu \Gamma^\mu{}_{00} + \Gamma^\lambda{}_{0\mu}\Gamma^\mu{}_{\lambda 0} - \Gamma^\lambda{}_{00}\Gamma^\mu{}_{\lambda\mu}. \tag{6.75}$$

(e) To deal with the $\Gamma\Gamma$-terms in part (d), write the familiar

$$\Gamma^\lambda{}_{\mu\nu} = \frac{1}{2}g^{\lambda\sigma}[\partial_\mu g_{\nu\sigma} + \partial_\nu g_{\mu\sigma} - \partial_\sigma g_{\mu\nu}]. \tag{6.76}$$

With $g_{\mu\nu} \approx \eta_{\mu\nu} + h_{\mu\nu}$ and $|h_{\mu\nu}|$ very small, show that

$$\Gamma^\lambda{}_{0\mu}\Gamma^\mu{}_{\lambda 0} - \Gamma^\lambda{}_{00}\Gamma^\mu{}_{\lambda\mu} \approx 0. \tag{6.77}$$

(f) Finally, show that $R_{00} \approx -\nabla^2\phi$ so that the surviving component of Einstein's field equations gives $A(-2\nabla^2\phi) = 8\pi G\rho$ and therefore, upon comparison with Poisson's equation, $A = -1$.

6.4 Show that Einstein's field equations (with $\Lambda = 0$),

$$R_{\mu\nu} - \frac{1}{2}g_{\mu\nu}R = -8\pi G T_{\mu\nu}, \tag{6.78}$$

can also be written

$$R_{\mu\nu} = -8\pi G\left(T_{\mu\nu} - \frac{1}{2}g_{\mu\nu}T^\lambda{}_\lambda\right). \tag{6.79}$$

6.5 From Ex. 6.4, we can see that, in a region devoid of all matter and other forms of energy, the stress-energy tensor vanishes and the field equations reduce to

$$R_{\mu\nu} = 0. \tag{6.80}$$

Consider a static point mass M sitting at the origin of a system of spherical coordinates. From the symmetry, the metric tensor components cannot depend on the coordinates t, θ, or φ. Therefore, making the minimum changes to the

Minkowski metric tensor, with the radial symmetry the metric tensor for the spacetime around M can be parameterized as

$$g_{\mu\nu} = \begin{pmatrix} A(r) & 0 & 0 & 0 \\ 0 & -B(r) & 0 & 0 \\ 0 & 0 & -r^2 & 0 \\ 0 & 0 & 0 & -r^2\sin^2\theta \end{pmatrix}. \tag{6.81}$$

Set up the differential equations that one must solve for $A(r)$ and $B(r)$. The solution to these equations is the "Schwarzschild solution," Eq. (3.53). The mass M enters the solution through the condition that $g_{00} \approx 1 + 2\phi$ (or $1 + 2\phi/c^2$ in conventional units) in the Newtonian limit. The strategy for solving the field equations is therefore to solve them in empty space "here" and impose the boundary condition that a point mass M "over there" contributes to g_{00} the potential $\phi = -GM/r$.

6.6 Starting from the flat-spacetime inhomogeneous Maxwell equations, $\partial_\mu F^{\mu\nu} = \mu_o j^\nu$, build the generally covariant version of them assuming the ABC metric of Ex. 4.4.

6.7 From Eq. (6.32) and the simplest Lorentz transformation, derive Eqs. (6.34) - (6.39), the expressions for the components of $\mathbf{E}'$ and $\mathbf{B}'$ in terms of $\mathbf{E}$, $\mathbf{B}$, and $\mathbf{v}_r$.

6.8 Show that the sources represented by j^μ and the potentials A^μ are the components of first-order tensors. Eq. (3.110) may offer a good place to start; see Charap, Chapter 4.

6.9 Within the context of the calculus of variations, consider the functional with one dependent variable,

$$J = \int_a^b L(t, x, \dot{x})dt. \tag{6.82}$$

The $x(t)$ that makes J a maximum or a minimum will satisfy the Euler-Lagrange equation,

$$\frac{\partial L}{\partial x} = \frac{d}{dt}\frac{\partial L}{\partial \dot{x}}. \tag{6.83}$$

Now consider a change of independent and dependent variables, $t' = T(t, x, \epsilon)$ and $x' = X(t, x, \epsilon)$, where ϵ is a continuous parameter, with $\epsilon = 0$ being the identity transformation. Under an infinitesimal transformation, to first order in ϵ the transformations become $t' \approx t + \epsilon\tau$ and $x' \approx x + \epsilon\zeta$, where $\tau = dT/d\epsilon|_0$ and $\zeta = dX/d\epsilon|_0$ are called the "generators" of the transformation. If the difference between the new functional and the old functional goes to zero faster than ϵ as $\epsilon \to 0$, and if the functional is also an extremal, then the conservation law of Emmy Noether's theorem results:

$$\frac{d}{dt}(p\zeta - H\tau) = 0, \tag{6.84}$$

where

$$\begin{aligned} p &\equiv \frac{\partial L}{\partial \dot{x}} \\ H &\equiv \dot{x}p - L = H(x,p,t) \end{aligned}$$

are the canonical momentum and the Hamiltonian, respectively. Generalize this procedure to fields, whose functional is a multiple integrals (e.g., over space-time), with n independent variables x^μ and fields $A^\mu = A^\mu(x^\nu)$. Thus, the functional will take the form

$$J = \int\int \cdots \int \mathcal{L}\left(x^\mu, A^\mu, \partial_\nu A^\mu\right)\, dx^{\mu_1} \cdots dx^{\mu_n}. \tag{6.85}$$

(a) Show that the Euler-Lagrange equation for the field theory becomes

$$\frac{\partial \mathcal{L}}{\partial A^\mu} = \partial_\lambda \left(\frac{\partial \mathcal{L}}{\partial (A^\mu{}_{,\lambda})} \right). \tag{6.86}$$

In this situation the canonical momenta and the Hamiltonian become tensors:

$$\begin{aligned} p^\lambda{}_\mu &= \left(\frac{\partial \mathcal{L}}{\partial (A^\mu{}_{,\lambda})} \right) \\ H^\mu{}_\nu &= (\partial_\lambda A^\mu) p^\lambda{}_\nu - \delta^\mu{}_\nu \mathcal{L}. \end{aligned}$$

Merely by using general covariance, show that Noether's conservation law becomes an equation of continuity,

$$\partial_\rho j^\rho = 0, \tag{6.87}$$

and identify j^ρ (see Neuenschwander).

Chapter 7

Tensors and Manifolds

Tensors model relationships between quantities in the physical world. Tensors are also logical relationships in the mental world of mathematical structures. Chapters 7 and 8 offer brief glimpses of tensors as mathematicians see them.

Particles and fields need space. The properties of the space itself are independent of coordinates. When coordinates are introduced, it is for the convenience of the problem solver, just as street addresses are introduced for the convenience of the traveler. Chapter 7 examines properties of space that transcend coordinates, such as the notion of a tangent space (which we have heretofore treated informally), and defines the continuous spaces called *manifolds*. Then the discussion will invoke coordinate systems and their basis vectors. We will see that many of the properties of tensor calculus, such as the definitions of the affine connection coefficients and covariant derivatives, can be expressed efficiently and effectively in terms of the basis vectors.

Chapter 8 takes us on a brief tour of tensors as seen in the context of abstract algebra, as mappings from a set of vectors to the real numbers. This elegant mathematical approach begins with the abstract notion of "multilinear forms" and wends its way to a very powerful tool called "differential forms" that incorporate a new kind of vector product, the "exterior product" denoted with a wedge, $\vec{a} \wedge \vec{b}$. Older books on general relativity (e.g., Einstein's *The Meaning of Relativity* and Eddington's *The Mathematical Theory of Relativity*) use tensors exclusively. Such references are still worth careful study and always will be. Many later books supplement tensors with the newer language of differential forms. When formally introduced as purely mathematical objects, differential forms can seem strange and unnecessary, raising barriers to their acceptance by physics students. Context and motivation are needed.

Let me offer a teaser. The vector identities

$$\nabla \times (\nabla \phi) = \mathbf{0} \tag{7.1}$$

and

$$\nabla \cdot (\nabla \times \mathbf{A}) = 0 \tag{7.2}$$

are special cases of one powerful identity, called the Poincaré lemma, that has this strange appearance:

$$\partial \wedge (\partial \wedge \psi) = 0, \tag{7.3}$$

alternatively denoted $\mathbf{d}(\mathbf{d}\psi) = 0$, where "$\partial \wedge \psi$" or "$\mathbf{d}\psi$" is a new kind of derivative called the "exterior derivative," and ψ is a member of a family of objects called "differential r-forms" where $r = 0, 1, 2, 3, \ldots$. If ψ is a "differential 0-form," then Eq. (7.3) reduces to Eq. (7.1). If ψ is a "differential 1-form," then Eq. (7.3) reduces to Eq. (7.2). To get to Poincaré's lemma, we need differential forms; to get differential forms, we must work out the r-forms that are *not* differential forms; to get to r-forms, we need to become familiar with the exterior product denoted by the wedge. This does not take as long as you might expect; the main barriers to understanding are jargon and motivation. The ideas themselves are no more complicated than anything else we have seen, but they seem strange at first because it is not clear to the novice *why* one should care about them.

Once we are on friendly terms with the exterior derivative, then line integrals, Gauss's divergence theorem, and Stokes's theorem will be seen as special cases of an integral theorem based on the antiderivative of Poincaré's lemma.

The introduction to differential forms encountered here will not make the initiate fluent in differential forms, but it should help make one conversational with them.

Notes on Notation

I need to specify the notation used in Chapters 7 and 8. Vectors in arbitrary spaces will henceforth be denoted with italicized names and carry an arrow above the name, such as $\vec{A}$. The components of vectors will be denoted with superscripts, such as A^μ. Basis vectors (not necessarily normalized to be *unit* vectors) are distinguished with subscripts, such as $\vec{e}_\mu$, to facilitate the summation convention. Thus, any vector can be expressed as a superposition over the basis vectors according to

$$\vec{A} = A^\mu \vec{e}_\mu. \tag{7.4}$$

Dual basis vectors are denoted with superscripts: $\vec{e}^\mu$. A set of basis vectors $\{\vec{e}^\mu\}$ dual to the set $\{\vec{e}_\nu\}$ is defined in terms of the latter according to

$$\vec{e}^\mu \cdot \vec{e}_\nu = \delta^\mu{}_\nu, \tag{7.5}$$

where $\delta^\mu{}_\nu$ denotes the Kronecker delta. A vector can be expanded as a superposition over the original basis, or over the dual basis. In the latter case the vector's components are denoted A_μ, and the superposition over the dual basis takes the form

$$\vec{A} = A_\mu \vec{e}^\mu, \tag{7.6}$$

where, again, repeated upper and lower indices are summed.

When vectors in three-dimensional Euclidean space are discussed, they will be presented in the boldface notation such as **A** familiar from introductory physics. Basis vectors that are normalized to unit length, the "unit vectors," will be denoted with hats, such as $\hat{e}_\mu$, $\hat{\mathbf{i}}$, $\hat{\boldsymbol{\rho}}$, and so on.

Second-order tensors will carry boldfaced names without arrows or tildes, such as the metric tensor **g**. Whether boldfaced symbols denote vectors in Euclidean 3-space or tensors more generally should be clear from the context.

In Chapter 8 the dual basis vectors will be re-envisioned as mathematical objects called "1-forms" and denoted with a tilde, $\tilde{A}$, with components A_μ. They can be written as superpositions over basis 1-forms denoted with tildes, such as $\tilde{\omega}^\mu$.

When r-forms (for $r > 1$) and their cousins, the differential forms, are introduced in Chapter 8, they will carry italicized symbols but no accent marks.

7.1 Tangent Spaces, Charts, and Manifolds

When one says a vector $\vec{A}$ exists at point P in a space, to allow for the possibility of curved space one means that the vector $\vec{A}$ resides in the tangent space T_P that exists at P. All statements about vectors and tensors are technically about what happens locally in the tangent space.

What is a tangent space? We have discussed it informally, exploiting the mental pictures of the derivative as the slope of the tangent line, the instantaneous velocity vector being tangent to a particle's trajectory, and a postage stamp Euclidean patch of surface that locally matches a curved surface. Outside of such infinitesimal regions, the tangent line or tangent vector or postage stamp does not necessarily coincide with the host curve or trajectory or surface.

Let n be the number of dimensions of a space. Any point P in the space, along with other *nearby* points, can be mapped locally to an n-dimensional Euclidean space. (In special and general relativity, any event P in spacetime, along with nearby events, can be mapped locally to a free-fall Minkowskian spacetime.) This notion expresses a familiar reality of cartography. Globally the Earth's surface is often modeled as the surface of a sphere. The paths of shortest distance between two points on the globe's surface, the geodesics, are great circles, the intersection of the globe's surface with a plane passing through its center. For all large triangles made of geodesics on the globe's surface, the sum of interior angles exceeds 180 degrees, demonstrating the non-Euclidean nature of this curved surface. But within small local regions, such as a university campus, the landscape can be adequately mapped on a sheet of paper, a flat Euclidean plane. Such a locally Euclidean coordinate grid offers the prototypical example of a "tangent space."

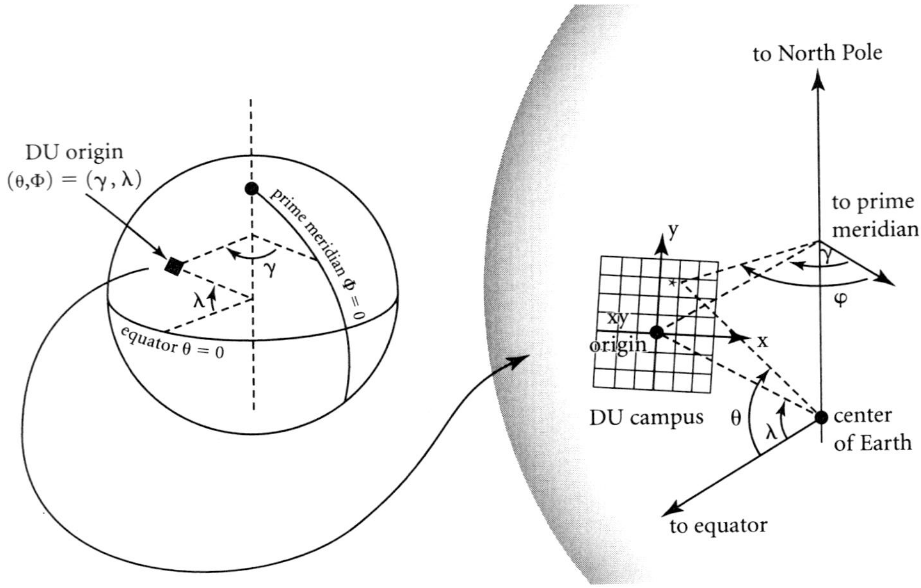

Figure 7.1: *Making a chart from a curved space to a locally flat space.*

Lay the local Euclidean map on the ground, and imagine extending its plane to infinity. Eventually the Earth curves out from under the extended plane, showing by this extrinsic procedure the map to be a tangent space, and the campus to be a small patch on the curved space. That is why a single Euclidean (x, y) grid cannot be extended to map the entire globe without distorting areas or angles. (If you have ever tried to wrap a basketball in a sheet of gift-wrapping paper, you see the problem.) For instance, this is why Greenland appears unrealistically huge on a Mercator projection. To have an accurate cartography of the *entire* globe with *Euclidean* maps requires an atlas of them. An atlas is a lot of local Euclidean (or Minkowskian) maps stitched together.

A mapping of a small region of a curved space onto a local Euclidean space of the same number of dimensions is called a *chart* by mathematicians. In our example of the globe the details work out as follows. If the Earth's surface is modeled as a sphere of radius a, any point on it can be mapped with two numbers, latitude and longitude. Let θ measure latitude (this θ equals zero at the equator, $+\pi/2$ at the North Pole, and $-\pi/2$ at the South Pole). Let φ measure longitude (zero at the prime meridian through Greenwich, measured positive going west). A student at Difficult University maps her campus with an xy coordinate system (x going east, y going north). Let the latitude and longitude of her xy chart's origin be λ and γ, respectively (see Figure 7.1). In terms of the global coordinates, a point on the campus will have local xy coordinates

$$x = a(\varphi - \gamma)$$
$$y = a(\theta - \lambda).$$

The x and y axes are straight, while the globe's surface is curved, so the chart and a patch of the globe's surface will agree only locally. Therefore, we acknowledge the crucial point that the angle differences $\varphi - \gamma$ and $\theta - \lambda$ *must be small* so that any distance between nearby points P and Q as measured on the chart agrees to high precision with the "actual" distance between P and Q as measured by following a geodesic along the globe.

Another observer at Challenging University maps his campus with an $x'y'$ coordinate system (x' east and y' north), with this system's origin at latitude λ' and longitude γ'. A point on his chart will have local $x'y'$ coordinates

$$x' = a(\varphi - \gamma')$$
$$y' = a(\theta - \lambda').$$

Since observers and instruments confined *to the surface* do not have a view from on high, a mental picture of the entire global surface will emerge for them only after enough local Euclidean charts have been made and laid side by side to make an atlas of the entire globe. To measure accurately *and intrinsically* the properties of the globe's surface, wherever the xy and $x'y'$ charts overlap, they must agree. Suppose the two campuses are close enough that the xy and $x'y'$ charts share a region of overlap. For instance, both charts might include the downtown district that stands between the two campuses. We can combine the two charts by overlaying them in their common overlap region, to give us a view of a part of the globe which stretches all the way from the first university to the second, a larger view than each chart provides alone. To map one chart directly to another, invert one of the transformations, for instance, by solving for the global coordinates that locate a point on CU's map,

$$\varphi = \frac{x'}{a} + \gamma'$$
$$\theta = \frac{y'}{a} + \lambda',$$

and then insert these results into the DU mapping, to show where any point on CU's map would be plotted on DU's map:

$$x = x' + a(\gamma' - \gamma)$$
$$y = y' + a(\lambda' - \lambda).$$

This inversion of one locally Euclidean chart, with the aim of relating it to another locally Euclidian chart, offers an example of what mathematicians call *transition maps*. When enough local Euclidean maps or charts of the local tangent spaces have been made and compared through their overlaps, such that the entire space has been consistently mapped, one says that an *atlas* of the entire space has been assembled. If the atlas is connected (has no discontinuities), then the global n-dimensional space is said to be a *manifold*.

The examples of manifolds most familiar to physicsts include coordinate grids that map Euclidean spaces and the four-dimensional spacetimes of special and general relativity. When we think about it, other manifolds come to mind. A particle moving through three-dimensional ordinary space sweeps out a trajectory through the six-dimensional manifold whose axes are labeled with possible values of the components of position and momentum. Such a manifold is called "phase space." A set of thermodynamic variables such as pressure P, temperature T, and volume V of an ideal gas can also be represented on manifolds. An equation of state, such as the well-known $PV = nRT$ for an ideal gas, defines a surface in the manifold. The terms "space" and "manifold" are sometimes used interchangeably, but to be precise, "manifold" means a continuous, differentiable, mathematical space such that, at every point P within it, a local Euclidean space of the same dimension can be constructed, the tangent space, whose chart coincides with the larger space in the neighborhood of P.

Because they are displacements from point a to point b, vectors are guaranteed to exist *only* in the tangent spaces of a manifold, but not necessarily across a finite section of the manifold itself. The importance of this point must not be underappreciated. This definition of a vector in terms of tangent space will be made precise below.

We did not have to worry about the distinction between the global space and a local tangent space in the Euclidean spaces of introductory mechanics or electrodynamics, or in the four-dimensional Minkowskian spacetime of special relativity. Those spaces or spacetimes were three- or four-dimensional "flat" manifolds. We may have used curvilinear *coordinates* (e.g., spherical coordinates), but the *space itself* was globally Euclidean or Minkowskian. A given tangent space's coordinate grid could be extended to infinity and accurately mapped the global coordinate grid because $g_{\mu\nu} = \eta_{\mu\nu}$ everywhere. Only in that context could the location of a point P, with its coordinates (x, y, z), be identified with a *single* displacement vector extending *all the way* from the origin to P, the vector

$$\vec{r} = (x, y, z) - (0, 0, 0), \tag{7.7}$$

which could be as long as we wanted. This was the notion behind our comments in Chapters 1 and 2 stating that a vector is a displacement. Among manifolds in general, vectors are still defined in terms of displacements, but technically they are defined only as *infinitesimal displacements within the tangent space* (infinitesimal so they can fit within the tangent space). Let us look at the definition of a vector from this perspective.

Consider a curve $\mathcal{C}$ wending its way through a manifold (like the Trans-Siberian Railroad wending its way across Russia). Let locations on $\mathcal{C}$ be described by a parameter λ. This λ could be the arc length along $\mathcal{C}$ measured from some reference point (e.g., the distance along the track from the Moscow station); in spacetime λ could be the proper time as measured by the wristwatch strapped to an observer in free fall along world line $\mathcal{C}$. Whatever the meaning of λ, at any point P on $\mathcal{C}$ a tangent vector $\vec{u}$ may be defined in the local tangent space (i.e., on the local chart). This is done through a limiting process, analogous to how the velocity vector is defined in introductory treatments. If point P has the value λ of the parameter, and a nearby point Q on C carries the value $\lambda + d\lambda$, the tangent vector from P to Q is defined according to a limit:

$$\vec{u} \equiv \lim_{d\lambda \to 0} \frac{Q(\lambda + d\lambda) - P(\lambda)}{d\lambda}. \tag{7.8}$$

At each point P in the manifold, a set of basis vectors $\{\vec{e}_\mu\}$ can be constructed to serve the locally flat tangent space T_P. These basis vectors do not necessarily have to be unit vectors. Within the tangent space, the tangent vector $\vec{u}$ may be written as the superposition

$$\vec{u}(P) = u^\mu \vec{e}_\mu(P), \tag{7.9}$$

where its components are given in terms of a coordinate grid according to

$$u^\mu = \lim_{d\lambda \to 0} \frac{x^\mu(\lambda + d\lambda) - x^\mu(\lambda)}{d\lambda} \equiv \frac{dx^\mu}{d\lambda} \tag{7.10}$$

evaluated at P.

In this way of looking at vectors, a vector field is defined point by point throughout the manifold by giving a recipe for determining it locally as a tangent vector. The global vector field is described as an atlas of local vectors.

7.2 Metrics on Manifolds and Their Tangent Spaces

The location of a point P in the manifold is specified by its address. This can be done without coordinates. For instance, three points in space (or events in spacetime) could be identified with the names "Larry," "Moe," and "Curly." While unambiguous, this is not very useful when the location must be used in calculations or measurements of distances. Thus, the spaces of interest to physics are *metric spaces*. A space S becomes a metric space whenever a distance $s(P, Q)$, a real number, can be defined between any two points P and Q within S such that the definition satisfies these four conditions:

(1) $s(P, Q) \geq 0$ for all $P, Q \in S$;
(2) $s(P, Q) = 0$ if and only if P and Q are the same point;
(3) $s(P, Q) = s(Q, P)$ for all $P, Q \in S$;
(4) $s(P, Q) \leq s(P, R) + s(R, Q)$ for all $P, Q, R \in S$ (the "triangle inequality").

Distances do not need coordinates, but the measurement of distances and directions are facilitated by the introduction of coordinates. If point P resides at a location specified by the coordinates x^μ, and nearby point Q has coordinates

$x^\mu + dx^\mu$, then distance ds (or its square) between P and Q is defined by the specification of some rule,

$$ds^2 = s(x^\mu, dx^\mu), \tag{7.11}$$

which is a scalar. In the Euclidean plane, thanks to the theorem of Pythagoras, we have the familiar rule

$$ds^2 = dx^2 + dy^2. \tag{7.12}$$

But one may invent other metrics for the same plane, such as this example of a Finsler geometry,

$$ds^2 = \sqrt{dx^4 + dy^4}, \tag{7.13}$$

illustrating the point that metrics are *defined* quantities, equally legitimate so long as they satisfy the four criteria in the definition of distance. For example, in two-dimensional Euclidean space, suppose that $dx = 2$ and $dy = 2$. With the Pythagorean metric, $ds = \sqrt{4+4} = 2\sqrt{2}$, but the same displacement mapped with the Finsler metric gives $ds = \sqrt[4]{16+16} = 2\sqrt{\sqrt{2}}$. Any pair of points on the xy plane can have the distance defined between them by either Pythagoreas or Finsler or by some other rule, the lesson being that distance is a matter of *definition*, within the constraints (1)–(4) given above.

A Riemannian geometry, by definition, has a metric that turns coordinate displacements into distances through a quadratic expression of the form

$$ds^2 = g_{\mu\nu} dx^\mu dx^\nu. \tag{7.14}$$

This includes Pythagoras but excludes Finsler. Strictly speaking, the manifold is Riemannian if and only if all the $g_{\mu\nu}$ are nonnegative. Should one or more of the $g_{\mu\nu}$ coefficients be negative—as occurs in Minkowskian spacetime—the manifold is said to be pseudo-Riemannian.

We have postulated that the infinitesimal neighborhood about any point P in a manifold can be locally mapped to a tangent space that has a Euclidean (or Minkowskian) metric. If this assumption is valid, then in any Riemannian or pseudo-Riemannian manifold mapped with a global set of coordinates x^μ and having the metric tensor $g_{\mu\nu}$, one can always make a coordinate transformation $x^\mu \to x'^\mu$ such that, *in the neighborhood of a specific* P, $g^{\mu\nu}(x) \to g'^{\mu\nu}(x') = \eta_{\mu\nu}$. This is demonstrated with a Taylor series below. However, it is not possible, in general, to find a transformation such that $g'_{\mu\nu}(x') = \eta_{\mu\nu}$ *globally* because Riemannian and pseudo-Riemannian spaces are not necessarily globally flat.

In a Taylor series expansion of $g'_{\mu\nu}(x')$ about P, one finds

$$g'_{\mu\nu}(P) = \eta_{\mu\nu} + (x'^\rho - x'^\rho_P)(x'^\sigma - x'^\sigma_P)\left[\frac{\partial^2 g'_{\mu\nu}}{\partial x'^\rho \partial x'^\sigma}\right]_P + \cdots . \tag{7.15}$$

Notably missing in the series is the first derivative, at P, of the metric tensor in the transformed coordinates, because

$$\frac{\partial g'^{\mu\nu}}{\partial x'^\rho}|_P = 0. \tag{7.16}$$

The proof consists of showing that, given the symmetries involved, there are sufficient independent conditions for Eq. (7.16) to occur (see, e.g. Hartle, pp. 140-141, or Hobson, Efstathiou, and Lasenby, pp. 42-44; see also Ex. 7.4). In the neighborhood of P the metric is Euclidean (or Minkowskian) until second order in the coordinate differentials.

7.3 Dual Basis Vectors

We have seen how a vector is formally defined in the local tangent space T_P. Within each T_P a set of basis vectors $\{\vec{e}_\mu\}$ may be constructed. At P it is also convenient to construct an alternative set of basis vectors $\{\vec{e}^\mu\}$, which are dual to the set of $\vec{e}_\mu$. The dual basis vectors are defined from the original ones according to

$$\vec{e}^\nu \cdot \vec{e}_\mu \equiv \delta^\nu{}_\mu. \tag{7.17}$$

Notice that if $\vec{e}^\mu$ is not a *unit* basis vector, then neither is $\vec{e}_\mu$. A vector $\vec{A}$ can be written in two ways:

$$\vec{A} = A_\nu \vec{e}^\nu = A^\mu \vec{e}_\mu. \tag{7.18}$$

Thus, a displacement vector in the local tangent space, $d\vec{s} = \vec{u}d\lambda$, may be written as the superposition,

$$d\vec{s} = dx^\mu \vec{e}_\mu, \tag{7.19}$$

so that, conversely,

$$\vec{e}_\mu \equiv \lim_{\Delta x^\mu \to 0} \frac{\Delta \vec{s}}{\Delta x^\mu} \equiv \frac{\partial \vec{s}}{\partial x^\mu}, \tag{7.20}$$

which will be illustrated shortly with an example.

Notice also that

$$\begin{aligned} ds^2 &= d\vec{s} \cdot d\vec{s} \\ &= (dx^\mu \vec{e}_\mu) \cdot (dx^\nu \vec{e}_\nu) \\ &= (\vec{e}_\mu \cdot \vec{e}_\nu) dx^\mu dx^\nu \\ &= g_{\mu\nu} dx^\mu dx^\nu. \end{aligned}$$

Likewise, by using the dual coordinate basis, we may write

$$dx^\mu = d\vec{s} \cdot \vec{e}^\mu \tag{7.21}$$

and show from ds^2 that

$$g^{\mu\nu} = \vec{e}^\mu \cdot \vec{e}^\nu. \tag{7.22}$$

If the set of basis vectors are *unit* basis vectors (hats instead of arrows over them) in the tangent space T_P, then

$$\hat{e}_\mu \cdot \hat{e}_\nu = \eta_{\mu\nu} \tag{7.23}$$

and

$$\hat{e}^{\mu} \cdot \hat{e}^{\nu} = \eta^{\mu\nu}, \tag{7.24}$$

where $\eta_{\mu\nu} = \pm\delta_{\mu\nu}$, and similarly for $\eta^{\mu\nu}$.

Raising and Lowering Indices

The dual basis vectors can be used to write a scalar product $\vec{A} \cdot \vec{B}$ in four equivalent ways:

$$\vec{A} \cdot \vec{B} = (A^{\mu}\vec{e}_{\mu}) \cdot (B^{\nu}\vec{e}_{\nu}) = g_{\mu\nu}A^{\mu}B^{\nu}, \tag{7.25}$$

$$\vec{A} \cdot \vec{B} = (A_{\mu}\vec{e}^{\mu}) \cdot (B_{\nu}\vec{e}^{\nu}) = g^{\mu\nu}A_{\mu}B_{\nu}, \tag{7.26}$$

$$\vec{A} \cdot \vec{B} = (A^{\mu}\vec{e}_{\mu}) \cdot (B_{\nu}\vec{e}^{\nu}) = \delta_{\mu}{}^{\nu}A^{\mu}B_{\nu} = A^{\nu}B_{\nu}, \tag{7.27}$$

$$\vec{A} \cdot \vec{B} = (A_{\mu}\vec{e}^{\mu}) \cdot (B^{\nu}\vec{e}_{\nu}) = \delta^{\mu}{}_{\nu}A_{\mu}B^{\nu} = A_{\nu}B^{\nu}. \tag{7.28}$$

Comparing first and last results shows that

$$A_{\nu}B^{\nu} = (g_{\mu\nu}A^{\mu})B^{\nu}, \tag{7.29}$$

so that

$$A_{\nu} = g_{\mu\nu}A^{\mu}. \tag{7.30}$$

Likewise,

$$B^{\nu} = g^{\mu\nu}B_{\mu}. \tag{7.31}$$

The dual sets of basis vectors are related by the superpositions

$$\vec{e}_{\mu} = g_{\mu\nu}\vec{e}^{\nu} \tag{7.32}$$

$$\vec{e}^{\mu} = g^{\mu\nu}\vec{e}_{\nu}. \tag{7.33}$$

Transformation of Basis Vectors

When important results follow from several lines of reasoning, all ending up in the same place, the results are robust and the methods equivalent. Here some familiar results will be derived all over again, but this time with explicit reference to basis vectors. A given infinitesimal displacement in the tangent space describes the same vector whether we express it with the set of coordinates x^{μ} or with another set of coordinates x'^{μ}:

$$d\vec{s} = dx^{\mu}\vec{e}_{\mu} = dx'^{\mu}\vec{e'}_{\mu}. \tag{7.34}$$

Since each new coordinate is a function of all the old coordinates, $x'^{\mu} = x'^{\mu}(x^{\nu})$, and vice versa $x^{\mu} = x^{\mu}(x'^{\nu})$, by the chain rule the coordinate differentials transform as

$$dx^{\mu} = \frac{\partial x^{\mu}}{\partial x'^{\nu}}dx'^{\nu}. \tag{7.35}$$

Therefore, Eq. (7.34) becomes

$$dx'^{\nu}\left(\vec{e'}_{\nu} - \frac{\partial x^{\mu}}{\partial x'^{\nu}}\vec{e}_{\mu}\right) = 0, \tag{7.36}$$

and thus each new basis vector is a superposition of all the old ones,

$$\vec{e'}_{\nu} = \frac{\partial x^{\mu}}{\partial x'^{\nu}}\vec{e}_{\mu}. \tag{7.37}$$

Similarly, it follows that

$$\vec{e'}^{\nu} = \frac{\partial x^{\mu}}{\partial x'^{\nu}}\vec{e}^{\mu}. \tag{7.38}$$

Furthermore, since

$$A^{\mu} = \vec{e}^{\mu} \cdot \vec{A}, \tag{7.39}$$

we find the transformation rule for vector components:

$$\begin{aligned} A'^{\mu} &= \vec{e'}^{\mu} \cdot \vec{A} \\ &= \frac{\partial x'^{\mu}}{\partial x^{\nu}}\vec{e}^{\nu} \cdot \vec{A} \\ &= \frac{\partial x'^{\mu}}{\partial x^{\nu}}A^{\nu}. \end{aligned}$$

Similarly,

$$A'_{\mu} = \frac{\partial x^{\nu}}{\partial x'^{\mu}}A_{\nu}. \tag{7.40}$$

Notice that the transformation of the basis vectors generates a *new vector* from all of the old ones, whereas transformation of vector components keeps the *original vector* but rewrites its components in the new coordinate system.

For an example of using these basis vector manipulations, consider the Euclidean plane mapped with rectangular (x, y) or with polar (ρ, θ) coordinates. A displacement can be written

$$d\vec{s} = (dx)\hat{\mathbf{i}} + (dy)\hat{\mathbf{j}} \tag{7.41}$$

in xy coordinates, and as

$$d\vec{s} = (d\rho)\vec{e}_{\rho} + (\rho d\theta)\vec{e}_{\theta} \tag{7.42}$$

in ρ, θ coordinates. The two systems are related by

$$\begin{aligned} x &= \rho\cos\theta \\ y &= \rho\sin\theta. \end{aligned}$$

The transformation rule of Eq. (7.37) says

$$\begin{aligned} \vec{e}_{\rho} &= \frac{\partial x}{\partial \rho}\hat{\mathbf{i}} + \frac{\partial y}{\partial \rho}\hat{\mathbf{j}} \\ \vec{e}_{\theta} &= \frac{\partial x}{\partial \theta}\hat{\mathbf{i}} + \frac{\partial y}{\partial \theta}\hat{\mathbf{j}}, \end{aligned}$$

which gives

$$\vec{e}_\rho = \hat{\mathbf{i}}\cos\theta + \hat{\mathbf{j}}\sin\theta \tag{7.43}$$

$$\vec{e}_\theta = \rho(-\hat{\mathbf{i}}\sin\theta + \hat{\mathbf{j}}\cos\theta). \tag{7.44}$$

The metric tensor components, $g_{\mu\nu} = \vec{e}_\mu \cdot \vec{e}_\nu$, are found to be

$$\begin{aligned} g_{\mu\nu} &= \begin{pmatrix} g_{\rho\rho} & g_{\rho\theta} \\ g_{\theta\rho} & g_{\theta\theta} \end{pmatrix} \\ &= \begin{pmatrix} 1 & 0 \\ 0 & \rho^2 \end{pmatrix}, \end{aligned}$$

and thus

$$\begin{aligned} d\vec{s} \cdot d\vec{s} &= g_{\mu\nu} dx^\mu dx^\nu \\ &= d\rho^2 + \rho^2 d\theta^2 \end{aligned}$$

as expected.

From $g^{\mu\sigma} g_{\nu\sigma} = \delta^\mu{}_\nu$ one sets up a system of four equations for the four unknown $g^{\mu\nu}$ and finds

$$g^{\mu\nu} = \begin{pmatrix} 1 & 0 \\ 0 & \frac{1}{\rho^2} \end{pmatrix}. \tag{7.45}$$

Turning now to the dual vectors $\vec{e}^\mu = g^{\mu\nu}\vec{e}_\nu$, we find from Eq. (7.33)

$$\begin{aligned} \vec{e}^\rho &= g^{\rho\rho}\vec{e}_\rho + g^{\rho\theta}\vec{e}_\theta \\ &= \vec{e}_\rho \\ &= \hat{\mathbf{i}}\cos\theta + \hat{\mathbf{j}}\sin\theta \end{aligned}$$

and

$$\begin{aligned} \vec{e}^\theta &= g^{\theta\rho}\vec{e}_\rho + g^{\theta\theta}\vec{e}_\theta \\ &= \frac{1}{\rho^2}\vec{e}_\theta \\ &= \frac{1}{\rho}(-\hat{\mathbf{i}}\sin\theta + \hat{\mathbf{j}}\cos\theta). \end{aligned}$$

These basis vectors and their duals are easily confirmed to be orthogonal,

$$\vec{e}^\rho \cdot \vec{e}_\rho = 1 \quad \vec{e}^\rho \cdot \vec{e}_\theta = 0$$

$$\vec{e}^\theta \cdot \vec{e}_\rho = 0 \quad \vec{e}^\theta \cdot \vec{e}_\theta = 1.$$

7.4 Derivatives of Basis Vectors and the Affine Connection

A point P in a manifold has its local tangent space T_P, and a different point Q has its local tangent space T_Q. Recall how parallelogram addition was invoked in introductory physics to define vector addition. One vector had to be moved with its direction held fixed, joining the other vector tail-to-head so the resultant could be constructed. But if the vectors reside in two different tangent spaces, it may not be possible to parallel-transport a vector from T_P to T_Q. Only in flat spaces with no curvature, as in globally Euclidean spaces and globally Minkowskian spacetimes, will T_P and T_Q join smoothly together in a universal tangent space. Since a derivative is defined by comparing the quantity at two nearby points, it follows that derivatives of basis vectors can be defined, in general, only locally in tangent spaces.

Of course, the n-dimensional curved manifold can be embedded in a Euclidean space (or Minkowskian spacetime) of $n + 1$ dimensions, analogous to how a globe's surface can be studied by embedding it in three-dimensional Euclidean space. In the $(n+1)$-dimensional emdedding space, tangent vectors from different parts of the globe *can* be compared, because parallel transport can be done over an arbitrarily large interval. In that embedding space, we may write

$$\Delta\vec{e}_\mu \equiv \vec{e}_\mu(Q) - \vec{e}_\mu(P), \tag{7.46}$$

which defines the increment $\Delta\vec{e}_\mu$, where $\mu = 1, 2, \ldots, n+1$. With it the derivative of a basis vector may be defined through the usual limiting process:

$$\frac{\partial\vec{e}_\mu}{\partial x^\nu} \equiv \lim_{\Delta x^\mu \to 0} \frac{\vec{e}_\mu(x^\nu + \Delta x^\nu) - \vec{e}_\mu(x^\mu)}{\Delta x^\nu}, \tag{7.47}$$

which exists in the $(n+1)$-dimensional embedding space. To translate it back into the original manifold, this derivative must be projected from the embedding space onto the n-dimensional tangent space T_P at P. This projected limit formally defines the derivative of the basis vector with respect to a coordinate,

$$\frac{\partial\vec{e}_\mu}{\partial x^\nu} = \lim_{\Delta x^\nu \to 0} \left(\frac{\Delta\vec{e}_\mu}{\Delta x^\nu}\right)_{\|toT_P}. \tag{7.48}$$

This result will be a superposition of the local basis vectors within the tangent space T_P, with some coefficients ${\Gamma^\lambda}_{\mu\nu}$:

$$\partial_\nu\vec{e}_\mu = {\Gamma^\lambda}_{\mu\nu}\vec{e}_\lambda. \tag{7.49}$$

Anticipating with our notation a result yet to be confirmed, these coefficients ${\Gamma^\lambda}_{\mu\nu}$ at P turn out to be those of the affine connection, which will now be demonstrated.

To isolate the $\Gamma^{\lambda}{}_{\mu\nu}$, multiply Eq. (7.49) by $\vec{e}^{\rho}$ and then use $\vec{e}^{\rho}\cdot\vec{e}_{\lambda} = \delta^{\rho}{}_{\lambda}$:

$$\begin{aligned}\vec{e}^{\rho}\cdot(\partial_\nu \vec{e}_\mu) &= \Gamma^{\lambda}{}_{\mu\nu}(\vec{e}^{\rho}\cdot\vec{e}_{\lambda}) \\ &= \Gamma^{\lambda}{}_{\mu\nu}\delta^{\rho}{}_{\lambda} \\ &= \Gamma^{\rho}{}_{\mu\nu}.\end{aligned}$$

Alternatively, by evaluating

$$\partial_\rho(\vec{e}^{\mu}\cdot\vec{e}_\nu) = \partial_\rho(\delta^{\mu}{}_{\nu}) = 0, \tag{7.50}$$

it follows that

$$\vec{e}_\nu\cdot(\partial_\rho \vec{e}^{\mu}) = -\Gamma^{\mu}{}_{\rho\nu}. \tag{7.51}$$

The transformation of the $\Gamma^{\lambda}{}_{\mu\nu}$ under a change of coordinates follows from the transformation of the basis vectors, since

$$\begin{aligned}\Gamma'^{\lambda}{}_{\mu\nu} &= \vec{e'}^{\lambda}\cdot(\partial_{\mu'}\vec{e'}_\nu) \\ &= \left(\frac{\partial x'^{\lambda}}{\partial x^{\rho}}\vec{e}^{\rho}\right)\cdot\frac{\partial}{\partial x'^{\mu}}\left[\frac{\partial x^{\sigma}}{\partial x'^{\nu}}\vec{e}_\sigma\right] \\ &= \left(\frac{\partial x'^{\lambda}}{\partial x^{\rho}}\vec{e}^{\rho}\right)\cdot\left[\frac{\partial x^{\sigma}}{\partial x'^{\nu}}\frac{\partial \vec{e}_\sigma}{\partial x'^{\mu}} + \frac{\partial^2 x^{\sigma}}{\partial x'^{\mu}\partial x'^{\nu}}\vec{e}_\sigma\right] \\ &= \left(\frac{\partial x'^{\lambda}}{\partial x^{\rho}}\frac{\partial x^{\sigma}}{\partial x'^{\nu}}\vec{e}^{\rho}\right)\cdot[\partial_{\mu'}\vec{e}_\sigma] + \left(\frac{\partial x'^{\lambda}}{\partial x^{\rho}}\frac{\partial^2 x^{\sigma}}{\partial x'^{\mu}\partial x'^{\nu}}\right)[\vec{e}^{\rho}\cdot\vec{e}_\sigma].\end{aligned}$$

In the second term we recognize $\vec{e}^{\rho}\cdot\vec{e}_\sigma = \delta^{\rho}{}_{\sigma}$. In the first term, expand $\vec{e}^{\rho}\cdot(\partial_{\mu'}\vec{e}_\sigma)$ using the chain rule and Eq. (7.49),

$$\begin{aligned}\vec{e}^{\rho}\cdot(\partial_{\mu'}\vec{e}_\sigma) &= \vec{e}^{\rho}\cdot\left[\frac{\partial \vec{e}_\sigma}{\partial x^{\tau}}\frac{\partial x^{\tau}}{\partial x'^{\mu}}\right] \\ &= \vec{e}^{\rho}\cdot\left[\vec{e}_\lambda\Gamma^{\lambda}{}_{\sigma\tau}\right]\frac{\partial x^{\tau}}{\partial x'^{\mu}} \\ &= g_{\rho\lambda}\Gamma^{\lambda}{}_{\sigma\tau}\frac{\partial x^{\tau}}{\partial x'^{\mu}} \\ &= \Gamma^{\rho}{}_{\sigma\tau}\frac{\partial x^{\tau}}{\partial x'^{\mu}}.\end{aligned}$$

Put everything back together to obtain

$$\Gamma'^{\lambda}{}_{\mu\nu} = \frac{\partial x'^{\lambda}}{\partial x^{\rho}}\frac{\partial x^{\sigma}}{\partial x'^{\nu}}\frac{\partial x^{\tau}}{\partial x'^{\mu}}\Gamma^{\rho}{}_{\sigma\tau} + \frac{\partial^2 x^{\rho}}{\partial x'^{\mu}\partial x'^{\nu}}\frac{\partial x'^{\lambda}}{\partial x^{\rho}}. \tag{7.52}$$

This is the same result we obtained in Chapter 4 for the transformation of the affine connection coefficients, which was done there without the help of basis vectors. Thus, the $\Gamma^{\lambda}{}_{\mu\nu}$ defined there and the $\Gamma^{\lambda}{}_{\mu\nu}$ defined here at least *transform* the same. We next demonstrate that they *are* the same.

Start with

$$g_{\mu\nu} = \vec{e}_\mu\cdot\vec{e}_\nu \tag{7.53}$$

and evaluate its derivative with respect to a coordinate:

$$\begin{aligned}\partial_\rho g_{\mu\nu} &= (\partial_\rho \vec{e}_\mu)\cdot\vec{e}_\nu + \vec{e}_\mu\cdot(\partial_\rho\vec{e}_\nu)\\ &= (\Gamma^\lambda{}_{\rho\mu}\vec{e}_\lambda)\cdot\vec{e}_\nu + \vec{e}_\mu\cdot(\Gamma^\lambda{}_{\rho\nu}\vec{e}_\lambda)\\ &= \Gamma^\lambda{}_{\rho\mu}g_{\lambda\nu} + \Gamma^\lambda{}_{\rho\nu}g_{\lambda\mu}.\end{aligned}$$

Now rewrite this equation two more times, each time permuting the indices $\rho\mu\nu$, to obtain

$$\partial_\mu g_{\nu\rho} + \partial_\nu g_{\rho\mu} - \partial_\rho g_{\mu\nu} = 2g_{\lambda\rho}\Gamma^\lambda{}_{\mu\nu}. \tag{7.54}$$

Multiply this by $g^{\rho\sigma}$, and recall that $g^{\rho\sigma}g_{\lambda\rho} = \delta^\sigma{}_\lambda$, to rederive a result identical to that found in Chapter 4:

$$\Gamma^\sigma{}_{\mu\nu} = \frac{1}{2}g^{\rho\sigma}[\partial_\mu g_{\rho\nu} + \partial_\nu g_{\rho\mu} - \partial_\rho g_{\mu\nu}]. \tag{7.55}$$

Now we see that the affine connection coefficients of Chapter 4 are the same coefficients as the $\Gamma^\lambda{}_{\mu\nu}$ introduced in Eq. (7.49). The demonstration here highlights the role of basis vectors explicitly.

The covariant derivatives also follow immediately through the basis vectors. Start with $\vec{A} = A^\alpha\vec{e}_\alpha$ and differentiate with respect to a coordinate:

$$\begin{aligned}\partial_\mu\vec{A} &= (\partial_\mu A^\alpha)\vec{e}_\alpha + A^\alpha(\partial_\mu\vec{e}_\alpha)\\ &= (\partial_\mu A^\alpha)\vec{e}_\alpha + A^\alpha\Gamma^\lambda{}_{\mu\alpha}\vec{e}_\lambda\\ &= (\partial_\mu A^\lambda + A^\alpha\Gamma^\lambda{}_{\mu\alpha})\vec{e}_\lambda\\ &\equiv (D_\mu A^\lambda)\vec{e}_\lambda,\end{aligned}$$

where we identify the covariant derivative that we met before,

$$D_\mu A^\lambda \equiv \partial_\mu A^\lambda + \Gamma^\lambda{}_{\mu\alpha}A^\alpha. \tag{7.56}$$

Working with the dual basis $\vec{A} = A_\alpha\vec{e}^\alpha$, one derives in a similar manner

$$D_\nu A_\lambda = \partial_\nu A_\lambda - \Gamma^\mu{}_{\lambda\nu}A_\mu. \tag{7.57}$$

In contrast, for a scalar field φ,

$$D_\mu\varphi = \partial_\mu\varphi \tag{7.58}$$

because φ does not depend on the choice of basis vectors.

Returning to our example of mapping the Euclidean plane in rectangular and in polar coordinates, explicit forms for the $\Gamma^\lambda{}_{\mu\nu}$ follow in only a few lines of calculation from Eq. (7.51), where $x^1 = \rho$ and $x^2 = \theta$:

$$\Gamma^1{}_{11} = 0, \tag{7.59}$$

$$\Gamma^2{}_{11} = 0, \tag{7.60}$$

$$\Gamma^1{}_{21} = 0, \tag{7.61}$$

$$\Gamma^2{}_{21} = \frac{1}{\rho}, \tag{7.62}$$

$$\Gamma^1{}_{22} = -\rho, \tag{7.63}$$

$$\Gamma^2{}_{22} = 0. \tag{7.64}$$

By Eq. (7.56), the covariant divergence in plane polar coordinates takes the form

$$D_\mu A^\mu = \partial_\mu A^\mu + \Gamma^\mu{}_{\mu\nu} A^\nu, \tag{7.65}$$

where

$$\Gamma^\mu{}_{\mu\nu} = \Gamma^1{}_{1\nu} + \Gamma^2{}_{2\nu} \tag{7.66}$$

with $\Gamma^\mu{}_{\mu 1} = \frac{1}{\rho}$ and $\Gamma^\mu{}_{\mu 2} = 0$. Therefore,

$$\begin{aligned} D_\mu A^\mu &= (\partial_1 + \frac{1}{\rho})A^1 + \partial_2 A^2 \\ &= \frac{1}{\rho}\partial_\rho(\rho A^\rho) + \partial_\theta A^\theta. \end{aligned}$$

Rewrite this plane polar coordinate example in terms of the *unit* basis vectors,

$$\hat{\boldsymbol{\rho}} \equiv \frac{\vec{e}_\rho}{|\vec{e}_\rho|} = \vec{e}_\rho \tag{7.67}$$

$$\hat{\boldsymbol{\theta}} \equiv \frac{\vec{e}_\theta}{|\vec{e}_\theta|} = \frac{1}{\rho}\vec{e}_\theta. \tag{7.68}$$

Let $\tilde{A}^\mu$ temporarily denote a component of $\vec{A}$ in the unit basis vectors, which means

$$\vec{A} = \tilde{A}_\rho \hat{\boldsymbol{\rho}} + \tilde{A}_\theta \hat{\boldsymbol{\theta}}. \tag{7.69}$$

Therefore,

$$\tilde{A}^\rho = A^\rho \tag{7.70}$$

$$\tilde{A}^\theta = \frac{1}{\rho}A^\theta, \tag{7.71}$$

and thus

$$D_\mu A^\mu = \frac{1}{\rho}\partial_\rho(\rho\tilde{A}^\rho) + \frac{1}{\rho}\partial_\theta \tilde{A}^\theta, \tag{7.72}$$

which is how the divergence, expressed in plane polar coordinates, is usually presented on the inside front cover of electrodynamics textbooks (without the tilde!).

We have seen in this chapter that some of the machinery of tensor calculus, such as the affine connection and covariant derivatives, can also be seen as artifacts of the properties of basis vectors. This should not be too surprising since indices imply coordinates and coordinates imply basis vectors. In the next chapter we will look at tensors from another perspective: mappings and

"r-forms" where r denotes a nonnegative integer. The r-forms will then be extended to something curious and powerful called "differential forms."

7.5 Discussion Questions and Exercises

Discussion Questions

Q7.1 Continuous mappings that have a continuous inverse function, and thereby preserve the topological properties of the space, are called "homeomorphisms." Must the mappings that define charts and transitions between them be homeomorphisms? What happens if they are not?

Q7.2 The Mercator projection is familiar to generations of schoolchildren as the map that hangs in the front of countless classrooms, making Greenland and Canada look far bigger than they really are. It maps the surface of the globe onto the xy coordinates of the map according to

$$\begin{aligned} x &= \frac{w}{2\pi}\varphi \\ y &= \frac{h}{2\pi}\ln\tan\left(\frac{\pi-\theta}{2}\right), \end{aligned}$$

where w is the map's width and h its height. The angle θ measures latitude on the globe, with $\theta = 0$ at the North Pole and $\theta = \pi$ at the South Pole. The angle φ denotes longitude measured from the prime meridian. Since the Mercator map is plotted on a Euclidean sheet of paper, why can't it serve as a chart in the context of that term's usage in discussing manifolds?

Q7.3 Consider thermodynamic state variables internal energy U, pressure P, volume V, temperature T, and entropy S. The second law of thermodynamics says that the entropy of an isolated system never decreases. Combined with the first law of thermodynamics this means

$$dS = \frac{1}{T}dU + \frac{P}{T}dV \geq 0. \tag{7.73}$$

Can we interpret entropy as a "distance" in an abstract space of thermodynamic variables and treat thermodynamics as geometry (see Weinhold)?

Q7.4 Recall the discussion in Section 7.2 about it being possible to find a coordinate transformation such that, at a specific point or event P in a Riemannian or pseudo-Riemannian manifold, $\partial_{\rho'} g'^{\mu\nu}|_P = 0$ (Eq. (7.16)). Is there an inconsistency here with the affine connection coefficients,

$$\Gamma^{\lambda}{}_{\mu\nu} = \frac{1}{2}g^{\lambda\rho}[\partial_\mu g_{\nu\lambda} + \partial_\nu g_{\mu\lambda} - \partial_\rho g_{\mu\nu}], \tag{7.74}$$

not all being zero? Explain.

Exercises

7.1 Find the $\vec{e}^{\mu}$ and the $\vec{e}_{\mu}$, the affine connection coefficients, and the covariant divergence of a vector **A** in three-dimensional Euclidean space mapped with spherical coordinates.

7.2 From the Schwarzschild metric (see Eq. (3.53)), find the $\vec{e}^{\mu}$ and the $\vec{e}_{\mu}$, the affine connection coefficients, and the covariant divergence of a vector whose components are A^{μ}.

7.3 *Covariant derivatives* and *"intrinsic derivatives"*: Typically, a vector field $\vec{A}$ is defined throughout a manifold, $\vec{A} = \vec{A}(x^{\mu})$. Examples include velocity fields in fluid flow, or electromagnetic and gravitational fields. However, sometimes a vector is defined only along a contour $\mathcal{C}$ in the manifold. Such an instance occurs when a particle moves along $\mathcal{C}$. Its instantaneous momentum and spin are defined only at points on $\mathcal{C}$, not throughout the entire manifold. Let s denote a parameter (such as distance along $\mathcal{C}$ from a reference point) that distinguishes one point on $\mathcal{C}$ from another. Study the rate of change along $\mathcal{C}$, by considering $d\vec{A}/ds$, where $\vec{A}(s) = A^{\mu}(s)\vec{e}_{\mu}(s)$:

$$\begin{aligned}\frac{d\vec{A}}{ds} &= \frac{dA^{\mu}}{ds}\vec{e}_{\mu} + A^{\mu}\frac{d\vec{e}_{\mu}}{ds} \\ &= \frac{dA^{\mu}}{ds}\vec{e}_{\mu} + A^{\mu}\frac{\partial\vec{e}_{\mu}}{\partial x^{\nu}}\frac{\partial x^{\nu}}{\partial s}.\end{aligned}$$

Using Eq. (7.49), this may be written

$$A^{\mu}\frac{d\vec{e}_{\mu}}{ds} = \frac{dA^{\mu}}{ds}\vec{e}_{\mu} + A^{\mu}\Gamma^{\lambda}{}_{\mu\nu}\vec{e}_{\lambda}\frac{\partial x^{\nu}}{\partial s}. \tag{7.75}$$

Show that this is equivalent to a superposition of covariant derivative components over the basis vectors,

$$\frac{d\vec{A}}{ds} = \frac{DA^{\rho}}{Ds}\vec{e}_{\rho}. \tag{7.76}$$

The derivative $d\vec{A}/ds$ is called the "intrinsic derivative." See Hobson, Efstathiou, and Lasenby, pp. 71-73 and 107-108, for further discussion.

7.4 Consider a two-dimensional spacetime mapped with time and space coordinates (t, r), having the spacetime interval

$$d\tau^2 = Adt^2 - Bdr^2 + 2Cdtdr, \tag{7.77}$$

where A, B, and C are functions of t and r. In matrix representation,

$$g_{\mu\nu} = \begin{pmatrix} A & C \\ C & -B \end{pmatrix}. \tag{7.78}$$

Show that coordinate transformations exist such that, at least locally, this off-diagonal metric can be diagonalized and then rescaled into a Minkowski metric. This illustrates by explict construction the theorem that shows, to first order in $x'^{\mu} - x'^{\mu}_{P}$ about point P, that $g_{\mu\nu} = \eta_{\mu\nu}$ (recall Eqs. (7.15) and (7.16) in Sec. 7.2).

(a) First, the diaganolization: show that an orthogonal transformation of the form

$$\Lambda = \begin{pmatrix} \cos\psi & \sin\psi \\ -\sin\psi & \cos\psi \end{pmatrix}, \tag{7.79}$$

used to produce $\mathbf{g}' = \Lambda^{\dagger}\mathbf{g}\Lambda$, will give

$$g'_{\mu\nu} = \begin{pmatrix} U & 0 \\ 0 & -V \end{pmatrix}, \tag{7.80}$$

provided that

$$\psi = -\frac{1}{2}\tan^{-1}\frac{C}{A+B}. \tag{7.81}$$

What are U and V in terms of A, B, and C?

(b) Now that we have $\mathbf{g}'$ so that $d\tau^2 = U dt'^2 - V dr'^2$, find a rescaling $t' \to t'', r' \to r''$ that makes $g''_{\mu\nu} = \eta_{\mu\nu}$ (this $g''_{\mu\nu}$ corresponds to the $g'_{\mu\nu}$ of Eq. (7.15)).

(c) Why does this procedure work only locally in general? In other words, why can't a transformation always be found that will convert any given metric tensor into $\eta_{\mu\nu}$ *globally*?

Chapter 8

Getting Acquainted with Differential Forms

If one takes a close look at Riemannian geometry as it is customarily developed by tensor methods one must seriously ask whether the geometric results cannot be obtained more cheaply by other machinery. –Harley Flanders

Recent decades have seen tensors supplemented with the use of *exterior products*, also called *multivectors*. An important special case is found in *differential forms*. While a thorough study of these topics goes beyond the scope of this book, I would like to offer an informal introduction to these neo-tensor vocabularies.

My goal for this chapter is modest: to suggest why these objects are relevant to physics, and help make them desirable and approachable for students to whom these topics are new. A motivation for inventing differential forms may be seen in the attempt to unify some differential vector identities that involve the gradient, divergence, and curl. Could it be that the vector identities in Euclidean space,

$$\mathbf{\nabla} \times (\mathbf{\nabla}\phi) = \mathbf{0} \tag{8.1}$$

and

$$\mathbf{\nabla} \cdot (\mathbf{\nabla} \times \mathbf{A}) = 0, \tag{8.2}$$

are merely special cases of a single equation? If so, then could line integrals, Stokes's theorem, and Gauss's divegence theorem be merely three cases of one idea? In the language of differential forms the answers to these questions are "yes." Eqs. (8.1) and (8.2) are instances of a strange-looking equation called "Poincaré's lemma" which looks like this:

$$\partial \wedge (\partial \wedge \psi) = 0. \tag{8.3}$$

In an alternative notation the lemma is also written like this:

$$\mathbf{d}(\mathbf{d}\psi) = 0. \tag{8.4}$$

In the context of Poincaré's lemma, the meaning of ψ distinguishes Eq. (8.1) from Eq. (8.2). In Eq. (8.1), ψ is something called a "differential 0-form," but in Eq. (8.2), ψ is something else called a "differential 1-form." If this unification intrigues you, read on, as we meet these differential forms. But of course we must first discuss some preliminary concepts, including the notion of multilinear forms, and a new kind of vector product.

8.1 Tensors as Multilinear Forms

More precisely, the metric tensor $\mathbf{g}$ *is a machine with two slots for inserting vectors,* $\mathbf{g}(\ ,\ )$*. Upon insertion, the machine spews out a real number.*
–Misner, Thorne, and Wheeler

From the perspective of abstract algebra, a tensor is a mapping from a set of vectors to the real numbers. Metaphorically, a tensor of order N is like a machine with N slots. One inserts a vector into each of the slots. The machine chugs and grinds, and out pops a real number. For example, the second-order metric tensor is (metaphorically) a machine called $\mathbf{g}(\ ,\ )$, with two input slots for inserting vectors. Insert vectors $\vec{A}$ and $\vec{B}$ into the slots. The metric tensor generates the number

$$\mathbf{g}(\vec{A}, \vec{B}) \equiv \vec{A} \cdot \vec{B}, \tag{8.5}$$

as illustrated in Fig. 8.1. Linearity in the vector arguments is essential to the utility of this abstract tensor definition. For instance, if one inserts into one slot of $\mathbf{g}$ the vector $\alpha\vec{A} + \beta\vec{B}$, where α and β are scalars, and inserts the vector $\vec{C}$ into the other slot, the result is

$$\begin{aligned} \mathbf{g}(\alpha\vec{A} + \beta\vec{B}, \vec{C}) &= \alpha\mathbf{g}(\vec{A}, \vec{C}) + \beta\mathbf{g}(\vec{B}, \vec{C}) \\ &= \alpha\vec{A} \cdot \vec{C} + \beta\vec{B} \cdot \vec{C} \\ &= (\alpha\vec{A} + \beta\vec{B}) \cdot \vec{C}. \end{aligned}$$

As we saw in Chapter 7, a vector can be written in terms of a set of basis vectors, $\vec{A} = A^\mu \vec{e}_\mu$. Now the metric tensor "machine" gives the output

$$\begin{aligned} \mathbf{g}(\vec{A}, \vec{B}) &= \mathbf{g}(A^\mu \vec{e}_\mu, B^\nu \vec{e}_\nu) \\ &= A^\mu B^\nu \mathbf{g}(\vec{e}_\mu, \vec{e}_\nu) \\ &= g_{\mu\nu} A^\mu B^\nu, \end{aligned}$$

where

$$g_{\mu\nu} \equiv \mathbf{g}(\vec{e}_\mu, \vec{e}_\nu) = \vec{e}_\mu \cdot \vec{e}_\nu. \tag{8.6}$$

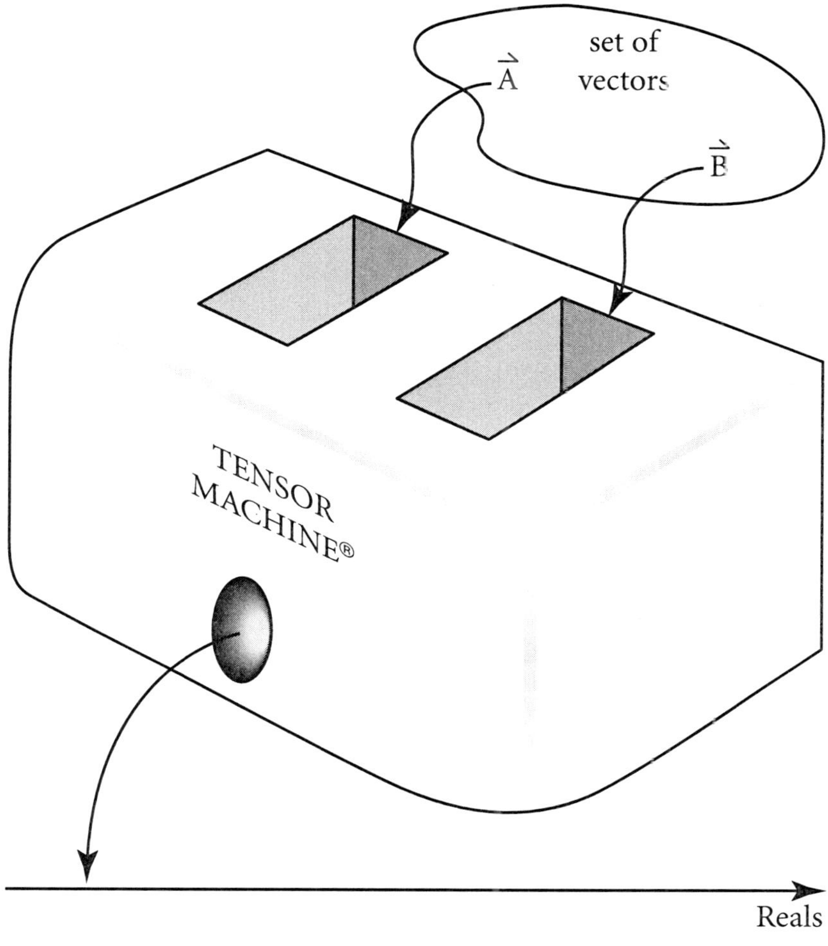

Figure 8.1: *A tensor of order N is like a machine with N input slots. Insert N vectors into the machine, one into each slot, and the machine generates a real number. Because the tensor mapping is linear by definition, each vector put into a slot may be a superposition of vectors.*

The same vector can also be written in terms of the dual set of basis vectors,

$\vec{A} = A_\mu \vec{e}^\mu$. Now the metric tensor machine gives the output

$$\begin{aligned}\mathbf{g}(\vec{A}, \vec{B}) &= \mathbf{g}(A_\mu \vec{e}^\mu, B_\nu \vec{e}^\nu) \\ &= A_\mu B_\nu \mathbf{g}(\vec{e}^\mu, \vec{e}^\nu) \\ &= g^{\mu\nu} A_\mu B_\nu,\end{aligned}$$

where

$$g^{\mu\nu} \equiv \mathbf{g}(\vec{e}^\mu, \vec{e}^\nu) = \vec{e}^\mu \cdot \vec{e}^\nu. \tag{8.7}$$

The tensor can handle mixtures of basis vectors and duals, such as

$$\mathbf{g}(\vec{e}_\mu, \vec{e}^\nu) \equiv g_\mu{}^\nu = \vec{e}_\mu \cdot \vec{e}^\nu \tag{8.8}$$

and so on. To say that a tensor has order $N = a + b$, usually denoted

$$\begin{pmatrix} a \\ b \end{pmatrix}, \tag{8.9}$$

means that for a tensor $\mathbf{R}$, its components

$$R(\vec{e}^{\mu_1}, \vec{e}^{\mu_2}, \ldots \vec{e}^{\mu_a}, \vec{e}_{\nu_1}, \vec{e}_{\nu_2}, \ldots \vec{e}_{\nu_b}) = R^{\mu_1\mu_2\cdots\mu_a}{}_{\nu_1\nu_2\cdots\nu_b} \tag{8.10}$$

are determined by inserting a dual basis vectors and b basis vectors into the various input slots.

In this chapter we also recast the notion of dual vectors into the concept of "1-forms." After that we mention briefly their generalizations to r-forms, where $r = 0, 1, 2, \ldots$. A subsequent step comes with the absorption of a coordinate differential dx^μ into each basis vector to redefine the basis vectors as $\vec{dx}^\mu \equiv dx^\mu \vec{e}_\mu$ (no sum over μ). Superpositions of these are formed with tensor components as coefficients. Thereby does the extension of r-forms to "differential forms" get under way. Of them Steven Weinberg wrote, "The rather abstract and compact notation associated with this formalism has in recent years seriously impeded communcation between pure mathematicians and physicists" (Weinberg, p. 114). Our sections here on r-forms and their cousins, the differential forms, aim merely to introduce these creatures, to enhance appreciation of the motivation behind their invention, and to glimpse why they are useful. I leave it for others to expound on the details and nuances of r-forms and differential forms. If I can help make these subjects interesting and approachable to physics students, that will be enough here.

Bilinear and Multilinear Forms

The dot product and thus the metric tensor offer an example of what mathematicians call a "bilinear form." Generically, in abstract algebra a bilinear form is a function $\Phi(\ ,\ )$ of variables $a, b, c, \ldots$ taken two at a time from a well-defined domain. Φ is linear in both input slots and outputs a real or complex number. By linearity, for a scalar α,

$$\Phi(\alpha a + c, b) = \alpha\Phi(a, b) + \Phi(c, b),$$
$$\Phi(a, \alpha b + c) = \alpha\Phi(a, b) + \Phi(a, c),$$

and so on.

This notion can be generalized to "multilinear forms," a function of N variables (thus N input slots), linear in each one:

$$\Phi(\ldots, \alpha a + \beta b, \ldots, c, \ldots) = \alpha\Phi(\ldots, a, \ldots, c, \ldots) + \beta\Phi(\ldots, b, \ldots, c, \ldots). \quad (8.11)$$

In the language of abstract algebra, a tensor is a multilinear form whose inputs are vectors and whose output is a real number. The number of slots is the order of the tensor. By way of concrete illustration, consider a tensor $\mathbf{T}$ of order 3, $\mathbf{T}(\ ,\ ,\ )$, into which three vectors, expanded over a basis, are inserted:

$$\begin{aligned}\mathbf{T}(\vec{A}, \vec{B}, \vec{C}) &= \mathbf{T}(A^\mu \vec{e}_\mu, B^\nu \vec{e}_\nu, C^\rho \vec{e}_\rho) \\ &= A^\mu B^\nu C^\rho \mathbf{T}(\vec{e}_\mu, \vec{e}_\nu, \vec{e}_\rho) \\ &= A^\mu B^\nu C^\rho T_{\mu\nu\rho}.\end{aligned}$$

Linearity is effectively utilized when vectors are written as a superposition of basis vectors.

Although tensors of order 2 or more are easily envisioned in terms of multilinear forms, from this perspective tensors of order 0 and order 1 ironically require more explanation. A scalar field $\phi(x^\mu)$ is a tensor of order 0 because it is a function of no vectors. This statement requires some reflection! In the study of electrostatics we encounter the electric potential $\phi(x, y, z)$. Its argument often gets abbreviated as $\phi = \phi(\mathbf{r})$, where $\mathbf{r} = (x, y, z) - (0, 0, 0)$ denotes the vector from the origin to the field point's location at (x, y, z). Such discourse on electrostatics gets away with saying $\phi = \phi(\mathbf{r})$ because the global Euclidean space and every point's tangent space are the same space. However, in generic manifolds, displacement vectors exist *only* locally in the tangent space, and thus the geometrical object $(x, y, z) - (0, 0, 0)$ may not reside within the manifold.

The point is this: just as the address "550 S. Eleventh Avenue" does not have to be the tip of a vector, likewise the coordinates x^μ merely specify the *address* that locates the point P uniquely, distinguishing it from all other locations in the manifold. We could have described equally well the location of P by naming it "Frodo" or by giving its distance along a specific contour $\mathcal{C}$ from some reference point on $\mathcal{C}$. The upshot is, scalar fields are functions of *location*, not, by definition, functions of *vectors*. A scalar, even a scalar field, is therefore a tensor of order 0. We can write $\phi(\mathbf{r})$ only because we are using $\mathbf{r}$ as the "address" when the distinction between a global space and its tangent spaces is not at issue.

To describe a tensor of order 1 as a mapping, we need a machine into which the insertion of one vector yields a real number. One approach would be to start with the second-order tensor $\mathbf{g}(\ ,\ )$ and insert one vector $\vec{A}$, leaving the other slot empty. This would generate a new machine $\tilde{A}(\ ) \equiv \mathbf{g}(\vec{A},\ )$, which can accommodate one real vector and yield a real number:

$$\tilde{A}(\vec{B}) = \mathbf{g}(\vec{A}, \vec{B}) = \vec{A} \cdot \vec{B}. \tag{8.12}$$

While $\tilde{A}(\ ) = \mathbf{g}(\vec{A},\ )$ offers an *example* of a 1-form, as a *definition* of a first-order tensor it would seem to depend on a preexisting second-order tensor, which seems backward! In the next section we examine 1-forms from another perspective and then generalize them to r-forms for $r = 0, 1, 2, \ldots$.

8.2 1-Forms and Their Extensions

A first-order tensor called a "1-form" takes a single vector $\vec{A}$ as input and cranks out a real number as output. Denote the 1-form as $\tilde{f}(\ )$, where the slot for the insertion of a vector has been indicated. Thus,

$$\tilde{f}(\vec{A}) = \alpha \tag{8.13}$$

for some real number α. What "machine" can be built within the manifold that produces one real number as output?

In the "standard model" visualization, in the local neighborhood of point P a 1-form $\tilde{f}$ is defined by specifying a "slicing of the space." Imagine a set of locally parallel planes. A scale for their spacing can be assigned. The orientation of the planes and their spacing defines the 1-form $\tilde{f}$ at P. When operating on a vector, by definition the 1-form at P counts the number of surfaces pierced by the vector there, as illustrated in Fig. 8.2. The vector $\vec{A}$ when put into $\tilde{f}(\ )$ yields the value 4; the vector $\vec{B}$ gives $\tilde{f}(\vec{B}) = 2.5$; and the vector $\vec{C}$ gives $\tilde{f}(\vec{C}) = 0$. When the vector is expanded as a superposition of basis vectors that have been erected in the tangent space at P, so that $\vec{A} = A^\mu \vec{e}_\mu$, then by the linearity of tensors,

$$\begin{aligned} \tilde{f}(\vec{A}) &= \tilde{f}(A^\mu \vec{e}_\mu) \\ &= A^\mu \tilde{f}(\vec{e}_\mu) \\ &= A^\mu f_\mu, \end{aligned}$$

which is a scalar (proved below), and where the coefficient

$$f_\mu = \tilde{f}(\vec{e}_\mu), \tag{8.14}$$

the μth component of the 1-form $\tilde{f}$, counts the number of the 1-form's surfaces pierced at P by $\vec{e}_\mu$.

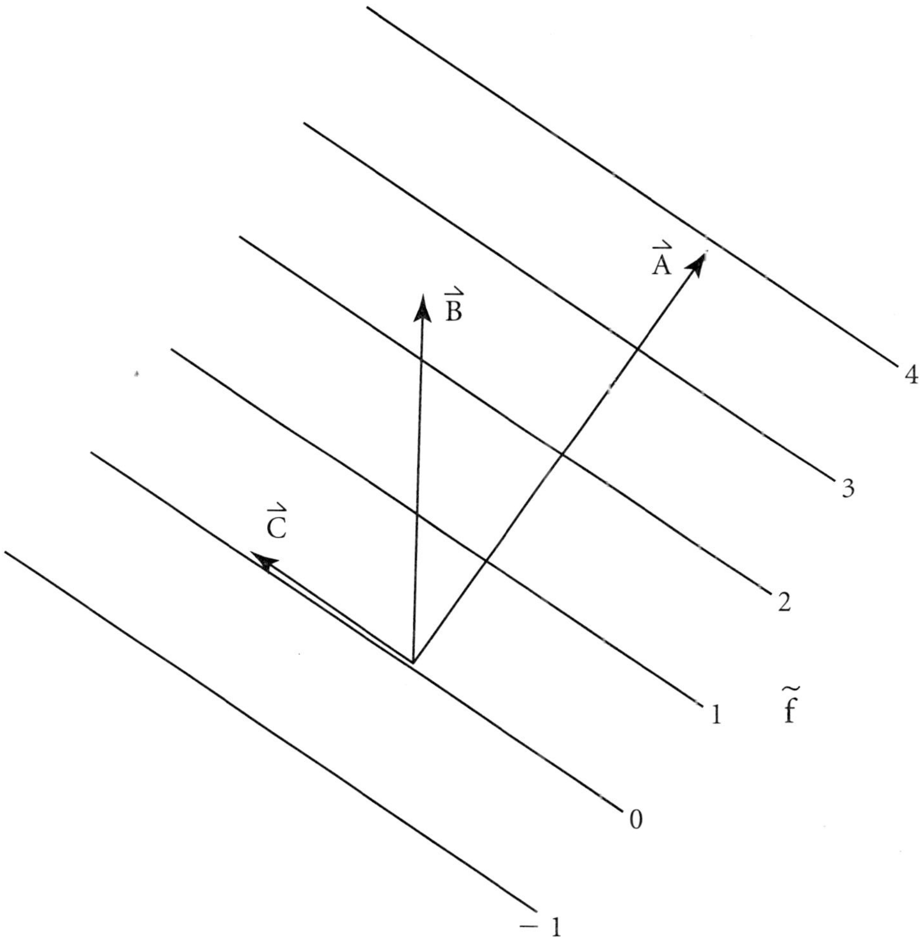

Figure 8.2: *Inserted into the 1-form $\tilde{f}(\;)$, the vectors $\vec{A}, \vec{B}$, and $\vec{C}$ give, respectively, the values 4, 2.5, and 0. The orientation of the locally parallel surfaces that "slice" the space, and their spacing, defines $\tilde{f}$. The number of surfaces (imagine as many interpolating surfaces as necessary) pierced by $\vec{A}$ is the real number $\tilde{f}(\vec{A})$.*

The scalar $\tilde{f}(\vec{A})$ is the fundamental scalar in Riemannian and psuedo-Riemannian geometries, of which the Euclidean 3-space $\mathbf{A}\cdot\mathbf{B}$ and the Minkowskian $dt^2 - ds^2$ are merely examples. It finds several notations in the literature:

$$\tilde{f}(\vec{A}) \equiv \langle \tilde{f}, \vec{A} \rangle \equiv \langle f|A \rangle, \tag{8.15}$$

all of which equal $f_\mu A^\mu$.

Now you are surely thinking, since a patch of surface can have a normal vector, why not just talk about the normal vector? To phrase this excellent question another way, could $\tilde{f}(\vec{A})$ merely be an elitist way of writing $\vec{f}\cdot\vec{A}$, where the vector $\vec{f}$ is normal to the locally parallel surfaces that define the 1-form $\tilde{f}$? Of course, in Euclidean space the suspected redundancy is correct: there a 1-form $\tilde{f}$ is indeed redundant with a vector $\vec{f}$, echoing how in such spaces the dual basis vectors $\vec{e}^{\mu}$ and the original basis vectors $\vec{e}_{\mu}$ are identical, rendering unnecessary the distinction between contravariant and covariant vector components.

But to keep the definition of the fundamental scalar the same across all Riemannian and pseudo-Riemannian geometries, the vector $|A\rangle$ and a dual $\langle B|$ must transform differently. We can see this from the transformation rule itself:

$$\begin{aligned} A'^{\mu}B'_{\mu} &= \left(\frac{\partial x'^{\mu}}{\partial x^{\rho}}A^{\rho}\right)\left(\frac{\partial x^{\sigma}}{\partial x'^{\mu}}B^{\sigma}\right) \\ &= \left(\frac{\partial x'^{\mu}}{\partial x^{\rho}}\frac{\partial x^{\sigma}}{\partial x'^{\mu}}\right)A^{\rho}B_{\sigma} \\ &= \delta^{\rho}{}_{\sigma}A^{\rho}B_{\sigma} \\ &= A^{\rho}B_{\sigma}. \end{aligned}$$

Since vectors transform differently from dual vectors, might it be misleading to represent them both as "arrows"? Different kinds of geometrical objects are entitled to different visual interpretations! Thus, the tradition has grown that "vectors are arrows" and "1-forms are surfaces." In the scalar product, the 1-form surfaces are pierced by a vector's arrow.

The set of 1-forms at P is closed under addition and scalar multiplication: If $\tilde{p}$ and $\tilde{q}$ are 1-forms and α is a scalar, then $\tilde{p}+\tilde{q}$ and $\alpha\tilde{p}$ are 1-forms at P. A set of 1-forms forms a "vector space" (also called a "linear space"; see Appendix C) in the sweeping abstract algebra sense of the term "vector." This means that, at any point P in a manifold of n dimensions, a set of basis 1-forms can be constructed, denoted $\tilde{\omega}^{\mu}$, where $\mu = 1, 2, \ldots n$. In terms of them any 1-form $\tilde{p}$ can be expressed as a superposition:

$$\tilde{p}(\vec{A}) = p_{\mu}\tilde{\omega}^{\mu}(\vec{A}), \tag{8.16}$$

where repeated indices are summed, and the p_{μ} are the components of $\tilde{p}$. To interpret $\tilde{\omega}^{\mu}(\vec{A})$, merely write $\vec{A}$ as a superposition of basis vectors:

$$\begin{aligned} \tilde{p}(\vec{A}) &= p_{\mu}\tilde{\omega}^{\mu}(A^{\nu}\vec{e}_{\nu}) \\ &= p_{\mu}A^{\nu}\tilde{\omega}^{\mu}(\vec{e}_{\nu}). \end{aligned}$$

Consistency between $\tilde{p}(\vec{A}) = p_{\mu}A^{\mu}$ and $\tilde{p}(\vec{A}) = p_{\mu}A^{\nu}\tilde{\omega}^{\mu}(\vec{e}_{\nu})$ requires that

$$\tilde{\omega}^{\mu}(\vec{e}_{\nu}) = \delta^{\mu}{}_{\nu}. \tag{8.17}$$

This becomes a set of equations for constructing the $\tilde{\omega}^{\mu}$ from given $\vec{e}_{\mu}$, as was similarly done in Chapter 7 with the dual basis vectors, where $\vec{e}^{\mu}\cdot\vec{e}_{\nu} = \delta^{\mu}{}_{\nu}$. Thus, the number of independent basis 1-forms equals the number of independent basis vectors, the dimensionality of the space.

Transformation of 1-forms

We saw in Chapter 7 how the coordinate independence of a vector's information content, expressed by

$$\vec{A} = A'^{\mu}\vec{e'}_{\mu} = A^{\mu}\vec{e}_{\mu}, \tag{8.18}$$

led to the transformation rule for the basis vectors, with each new basis vector a linear combination of all the old ones:

$$\vec{e'}_{\nu} = \frac{\partial x^{\mu}}{\partial x'^{\nu}}\vec{e}_{\mu}. \tag{8.19}$$

The corresponding vector components (same vector $\vec{A}$, projected into a new coordinate system) transform as

$$A'^{\nu} = \frac{\partial x'^{\nu}}{\partial x^{\mu}}A^{\mu}. \tag{8.20}$$

By similar reasoning, the transformation laws for basis 1-forms and their components are derived. Start with

$$\tilde{p} = p'_{\mu}\tilde{\omega}'^{\mu} = p_{\mu}\tilde{\omega}^{\mu}. \tag{8.21}$$

The notation p'_{μ} means that the 1-form $\tilde{p}$ (which has meaning independent of coordinates, just as a vector $\vec{A}$ transcends coordinates) operates on the basis vector $\vec{e'}_{\mu}$:

$$\begin{aligned} p'_{\mu} &= \tilde{p}(\vec{e'}_{\mu}) \\ &= \tilde{p}\left(\frac{\partial x^{\nu}}{\partial x'^{\mu}}\vec{e}_{\nu}\right) \\ &= \frac{\partial x^{\nu}}{\partial x'^{\mu}}\tilde{p}(\vec{e}_{\nu}) \\ &= \frac{\partial x^{\nu}}{\partial x'^{\mu}}p_{\nu}. \end{aligned}$$

This says that the *component* p_{μ} of a 1-form transforms the same as does the basis *vector* $\vec{e}_{\mu}$. For this reason the 1-forms are sometimes called "covectors," leading to the historical term "covariant vectors." Conversely, the components A^{μ} of the vectors transform opposite, or contrary, to the basis vectors $\vec{e}_{\mu}$, and for that reason they are said to be "contravariant" components. When 1-forms are envisioned as another kind of vector ("dual" to the original ones), these adjectives are necessary.

From $p_{\mu}\tilde{\omega}^{\mu} = p'_{\mu}\tilde{\omega}'^{\mu}$ and the transformation rule for the p_{μ}, the transformation of the basis 1-forms immediately follows,

$$\tilde{\omega}'^{\mu} = \frac{\partial x'^{\mu}}{\partial x^{\nu}}\tilde{\omega}^{\nu}. \tag{8.22}$$

All of this echoes what we have seen before: under a change of coordinate system, vector components transform one way,

$$A'^{\mu} = \frac{\partial x'^{\mu}}{\partial x^{\nu}} A^{\nu}, \tag{8.23}$$

and their dual or 1-form components transform another way,

$$A'_{\mu} = \frac{\partial x^{\nu}}{\partial x'^{\mu}} A_{\mu}. \tag{8.24}$$

It goes back to how vectors transform as displacements, whereas dual vectors or 1-forms transform as a quantity *per* displacement.

The Gradient as a 1-Form

We have seen how a 1-form maps a vector to a real number. The prototypical example emerges in the gradient. Let $\phi = \phi(x^{\mu})$ be a scalar field, a function of location as specified by coordinates. A change in ϕ corresponding to displacement dx^{ν} in the tangent space is given by

$$d\phi = (\partial_{\nu}\phi)dx^{\nu}. \tag{8.25}$$

Compare this to the result of a 1-form operating on the displacement vector $d\vec{s}$ that resides in the tangent space:

$$\begin{aligned} \tilde{p}(d\vec{s}) &= \tilde{p}(dx^{\nu}\vec{e}_{\nu}) \\ &= \tilde{p}(\vec{e}_{\nu})dx^{\nu} \\ &= p_{\nu}dx^{\nu}. \end{aligned}$$

Comparing $d\phi$ to $\tilde{p}(d\vec{s})$ suggests that $\partial_{\nu}\phi$ is the component of a 1-form. The 1-form itself is denoted

$$\tilde{d}\phi(\ ) \tag{8.26}$$

with an input slot, where

$$\tilde{d}\phi(\vec{e}_{\nu}) \equiv \partial_{\nu}\phi. \tag{8.27}$$

In this expression the basis vector $\vec{e}_{\nu}$ is the argument of the 1-form $\tilde{d}\phi$, not the argument of ϕ. In other words, $\tilde{d}\phi$ and $\tilde{d}\psi$ are different 1-forms, with components $\partial_{\nu}\phi$ and $\partial_{\nu}\psi$.

In three-dimensional Euclidean space mapped with xyz coordinates, the 1-form components are those of the familiar gradient, with the components

$$(\tilde{d}\phi)_i = \left(\frac{\partial\phi}{\partial x}, \frac{\partial\phi}{\partial y}, \frac{\partial\phi}{\partial z}\right) \equiv \nabla_i\phi. \tag{8.28}$$

In Minkowskian spacetime mapped with $txyz$ coordinates (note the necessity of a basis vector for the time direction, $\vec{e}_0$), the gradient 1-form components are

$$\begin{aligned} (\tilde{d}\phi)_{\mu} &= \left(\frac{\partial\phi}{\partial t}, \frac{\partial\phi}{\partial x}, \frac{\partial\phi}{\partial y}, \frac{\partial\phi}{\partial z}\right) \\ &= (\partial_t, \boldsymbol{\nabla})\phi. \end{aligned}$$

Components of the differential operator itself can be abstracted from the function ϕ,

$$(\tilde{d})_\mu = (\partial_t, \boldsymbol{\nabla}). \tag{8.29}$$

Whatever ϕ happens to be, the gradient 1-form $\tilde{d}\phi$ may operate on any vector. Writing $\vec{A}$ in terms of basis vectors gives

$$\begin{aligned}\tilde{d}\phi(\vec{A}) &= \tilde{d}\phi(A^\mu \vec{e}_\mu)\\ &= A^\mu \tilde{d}\phi(\vec{e}_\mu)\\ &= A^\mu \partial_\mu \phi.\end{aligned}$$

On the other hand, when the gradient is expanded in a 1-form basis, $\tilde{d}\phi = (\partial_\mu\phi)\tilde{\omega}^\mu$, then the same operation gets expressed as

$$\tilde{d}\phi(\vec{A}) = (\partial_\mu\phi)\tilde{\omega}^\mu(\vec{A}). \tag{8.30}$$

To proceed, we still have to expand $\vec{A}$ over a basis, which leads by this route to the result just obtained,

$$\begin{aligned}\tilde{d}\phi(\vec{A}) &= (\partial_\mu\phi)\tilde{\omega}^\mu(A^\nu \vec{e}_\nu)\\ &= (\partial_\mu\phi)A^\nu\tilde{\omega}^\mu(\vec{e}_\nu)\\ &= (\partial_\mu\phi)A^\nu\delta^\mu{}_\nu\\ &= (\partial_\nu\phi)A^\nu.\end{aligned}$$

The gradient as a 1-form becomes especially useful when the displacement vector $d\vec{s} = dx^\mu \vec{e}_\mu$ is inserted, for then

$$\tilde{d}\phi(d\vec{s}) = (\partial_\mu\phi)dx^\mu = d\phi, \tag{8.31}$$

which is a directional derivative in n dimensions. In Euclidean 3-space this reduces to a familiar result,

$$\tilde{d}\phi(d\vec{s}) = (\boldsymbol{\nabla}\phi)\cdot d\mathbf{s} = d\phi. \tag{8.32}$$

As Eq. (8.32) reminds us, in courses such as mechanics and electrodynamics we were taught to think of the gradient as a vector. It is fine to think of it this way in three-dimensional Euclidean space, where the 1-forms or dual vectors are redundant with vectors. It was through this notion of the directional derivative that the gradient acquired its vector interpretation in the first place (recall Section 1.4). This point of view requires the existence of the scalar product, and thus requires the metric tensor, as we see now:

$$\vec{\nabla}\phi \cdot d\vec{s} = \mathbf{g}(\vec{\nabla}\phi, d\vec{s}). \tag{8.33}$$

The claim that the gradient serves as the prototypical example of a 1-form, and the interpretation of a 1-form as a set of surfaces "slicing space," may seem more natural in light of a simple thought experiment from introductory physics.

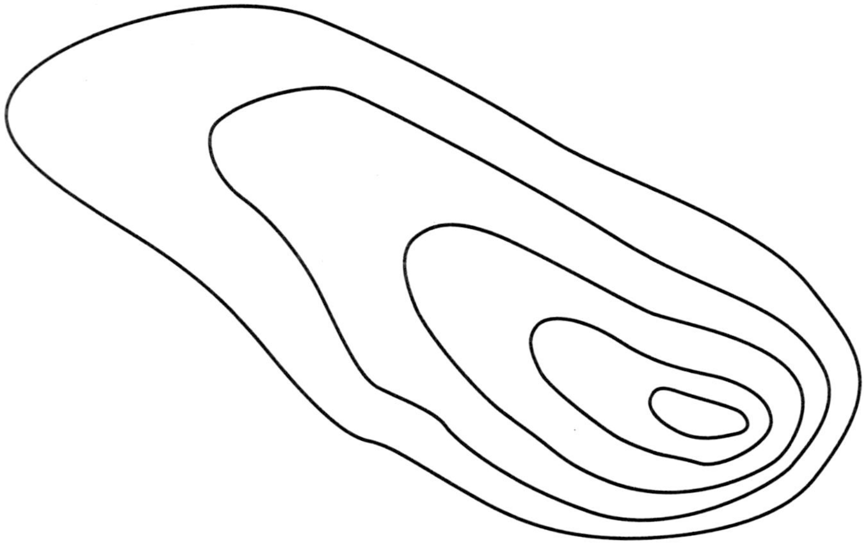

Figure 8.3: *A topographical map showing contours of constant elevation (constant potential energy) where the landscape topography intersects the horizontal equipotential planes.*

Let $U(z) = mgz$ represent the gravitational potential energy that describes the interaction between a particle of mass m and the Earth's gravitational field near the planet's surface where the field has uniform magnitude g. Here z denotes the vertical coordinate above a local xy plane on the Earth's surface. Equipotential surfaces form horizontal planes, each one at elevation z.

Imagine standing on the side of a mountain. The base of the mountain sits on the local xy coordinate grid. The mountain's visible shape brings to life a surface $\mathcal{S}$ above the xy plane. A topographical map (see Fig. 8.3) shows contour lines made by the intersection of equipotential surfaces with $\mathcal{S}$. Your altitude at a point on the mountain is $z = z(x, y)$. When you take a step, your displacement, as projected onto the xy plane, has components $dx^\mu = (dx, dy)$. Your altitude may go up, go down, or stay the same, depending on the direction of your step. Suppose you take a step that takes you upward toward the summit. In taking this step, you change your gravitational potential energy, cutting through some equipotential surfaces. If instead your step carries you along a line of constant elevation, then you cut through no equipotential surfaces. For a given displacement, the steeper the slope in the direction you go, the greater the number of equipotential surfaces that are pierced by your trajectory as you move through them. The frequency of spacing of the equipotential surfaces (say, surfaces 10 J apart) as you move along a path on the mountain's surface offers a visual representation of the 1-form. Where the slopes are gentle, many steps on $\mathcal{S}$ are required to change your potential energy by 10 J. In that region the 1-form, like the gradient, is small, which in turn means that $\Delta U = 10$ J contours are widely spaced on $\mathcal{S}$ in this vicinity. But where the slopes are steep, the $\Delta U = 10$ J contours on the topographical map are close together; it takes very few steps on $\mathcal{S}$ to cross enough equipotential surfaces for ΔU to equal 10 J, and the 1-form (steep gradient) is large in magnitude.

For a given displacement, piercing the equipotential surfaces at a high frequency means that the 1-form has a large magnitude. Piercing them at a low frequency means that the 1-form has a small magnitude. (See Misner, Thorne, and Wheeler, pp. 53-59; Schutz, *A First Course in General Relativity*, p. 66.)

Tensor Products and Tensors of Higher Order

New tensors are created from old ones of the same order by addition and scalar multiplication. Can new tensors of higher order be made from old ones by multiplication? In Section 2.7 we suggested that a component of a two-index tensor is merely the product of two vector components, third-order tensor components are products of three vector components, and so on. That notion generalizes to a formal algebraic operation called the *tensor product.* It works like this: Suppose U is a set of vectors in n-dimensional space, and V is another set of vectors in m-dimensional space. From these two vector spaces we define a new space, also linear, of nm dimensions. This new space is denoted $U \otimes V$. If $\vec{u}$ is an element of U and $\vec{v}$ an element of V, then the elements of $U \otimes V$ are denoted $\vec{u} \otimes \vec{v}$. These objects, whatever they are, respect the rules of linearity that are common to all abstract vector spaces:

$$\begin{aligned}
(\vec{u}_1 + \vec{u}_2) \otimes \vec{v} &= (\vec{u}_1 \otimes \vec{v}) + (\vec{u}_2 \otimes \vec{v}) \\
\vec{u} \otimes (\vec{v}_1 + \vec{v}_2) &= (\vec{u} \otimes \vec{v}_1) + (\vec{u} \otimes \vec{v}_2) \\
\alpha(\vec{u} \otimes \vec{v}) &= (\alpha\vec{u}) \otimes \vec{v} = \vec{u} \otimes (\alpha\vec{v}) \\
\vec{0} \otimes \vec{v} &= \vec{u} \otimes \vec{0} = 0.
\end{aligned}$$

What does $\vec{u}\otimes\vec{v}$ look like in terms of $\vec{u}$ and $\vec{v}$? Since the tensor product makes an nm-dimensional space out of an n-dimensional space and an m-dimensional space, if the elements of U are represented by row matrices and the elements of V are represented as column matrices, then the elements of $U \otimes V$ are rectangular matrices; for example, if $n = 2$ and $m = 3$, then

$$\begin{aligned}
\vec{u} \otimes \vec{v} &= \begin{bmatrix} a \\ b \end{bmatrix} \begin{bmatrix} r & s & t \end{bmatrix} \\
&= \begin{bmatrix} ar & as & at \\ br & bs & bt \end{bmatrix}.
\end{aligned}$$

For a particular example, the components Q^{jk} of the electric quadrupole tensor are $\int x^i x^j dq$, members of an object we may denote $\int \mathbf{r}\otimes\mathbf{r}\, dq$. I invite comparison here with the Dirac bracket notation used to illustrate the inertia tensor in Section 2.5.

Writing out $\vec{u} \otimes \vec{v}$ explicitly required us to put $\vec{u}$ and $\vec{v}$ in terms of components, which implies a coordinate system and basis vectors. The coordinate representation clearly shows that $\vec{u} \otimes \vec{v}$ forms an object more complicated than the vectors going into it. Although an explicit *representation* of $U \otimes V$ is facilitated by coordinates, the space $U \otimes V$ and its elements exist without coordinates. Vectors themselves, we recall, do not depend on coordinate systems for their existence. We anticipate that, if $n = m$, the components of $\vec{u} \otimes \vec{v}$ will be identified with the components of a second-order tensor. The distinction between upper indices and lower indices remains to be articulated in this language; we return to it below.

Once $\vec{u} \otimes \vec{v}$ has been defined, objects such as $\vec{u} \otimes \vec{v} \otimes \vec{w}$ can also be defined, a tensor product of three vectors, elements of a space denoted $U \otimes V \otimes W$ whose dimensionality is the product of the dimensions of the spaces U, V, and W. One may continue in this manner with tensor products of higher order.

If $\tilde{p}(\vec{A}) = \alpha$ and $\tilde{q}(\vec{B}) = \beta$, where α and β are real numbers, then a second-order tensor output can be thought of as a tensor product of two 1-forms,

$$\mathbf{T}(\vec{A}, \vec{B}) = \tilde{p}(\vec{A}) \otimes \tilde{q}(\vec{B}) = \alpha\beta. \tag{8.34}$$

This tensor product $\tilde{p} \otimes \tilde{q}$ is also called the "direct product" of $\tilde{p}$ and $\tilde{q}$. From this perspective, a second-order tensor is also called a "2-form." Expand the vectors in terms of a basis, and invoke linearity. Doing so puts us on notationally familiar ground:

$$\begin{aligned}
\tilde{p}(\vec{A}) \otimes \tilde{q}(\vec{B}) &= \mathbf{T}(\vec{A}, \vec{B}) \\
&= \mathbf{T}(A^\mu \vec{e}_\mu, B^\nu \vec{e}_\nu) \\
&= A^\mu B^\nu \mathbf{T}(e_\mu, \vec{e}_\nu) \\
&= A^\mu B^\nu T_{\mu\nu},
\end{aligned}$$

where

$$T_{\mu\nu} \equiv \mathbf{T}(\vec{e}_\mu, \vec{e}_\nu). \tag{8.35}$$

A set of basis 2-forms, denoted $\tilde{\omega}^{\mu\nu}$, may be envisioned so that any 2-form can be expressed as a superposition of them, according to

$$\mathbf{T}(\ ,\ ) = T_{\mu\nu}\tilde{\omega}^{\mu\nu}(\ ,\ ). \tag{8.36}$$

Specifically, since the order-2 tensor is a direct product of 1-forms, we can define (with input slots)

$$\tilde{\omega}^{\mu\nu}(\ ,\ ) = \tilde{\omega}^\mu(\ ) \otimes \tilde{\omega}^\nu(\ ). \tag{8.37}$$

Let us do a consistency check. On the one hand, $T_{\mu\nu} = \mathbf{T}(\vec{e}_\mu, \vec{e}_\nu)$. On the other hand, in terms of the second-order basis,

$$\begin{aligned}
\mathbf{T}(\vec{e}_\mu, \vec{e}_\nu) &= T_{\rho\sigma}\tilde{\omega}^{\rho\sigma}(\vec{e}_\mu, \vec{e}_\nu) \\
&= T_{\rho\sigma}\tilde{\omega}^{\rho}(\vec{e}_\mu) \otimes \tilde{\omega}^{\sigma}(\vec{e}_\nu).
\end{aligned}$$

Consistency will be obtained if and only if

$$\omega^{\rho\sigma}(\vec{e}_\mu, \vec{e}_\nu) = \delta^\rho{}_\mu \delta^\sigma{}_\nu, \tag{8.38}$$

analogous to Eq. (8.17). One continues in like manner to higher orders.

1-Forms as Mappings

A tensor has been defined as a mapping from a set of vectors to the real numbers. We can turn this around and think of a tensor as mapping a set of 1-forms to the real numbers:

$$\mathbf{T}(\tilde{p}, \tilde{q}) = \gamma \tag{8.39}$$

for some real number γ. Breaking the 1-forms down into a superposition over a basis set of 1-forms, we may write

$$\begin{aligned}\mathbf{T}(\tilde{p}, \tilde{q}) &= \mathbf{T}(p_\mu \tilde{\omega}^\mu, q_\nu \tilde{\omega}^\nu) \\ &= p_\mu q_\nu \mathbf{T}(\tilde{\omega}^\mu, \tilde{\omega}^\nu) \\ &= p_\mu q_\nu T^{\mu\nu} = \gamma,\end{aligned}$$

where the components $T^{\mu\nu}$ of a second-order contravariant tensor are the real numbers that emerge from the machine $\mathbf{T}(\ ,\ )$ operating on a pair of basis 1-forms:

$$T^{\mu\nu} \equiv \mathbf{T}(\tilde{\omega}^\mu, \tilde{\omega}^\nu). \tag{8.40}$$

One can generalize further and imagine a tensor of order $N = a + b$ to be like a machine into which one inserts a 1-forms and b vectors to generate a single real number. For instance, into a 4-slot tensor machine $\mathbf{R}$, if we insert one 1-form and three vectors, the real number that emerges from the output will be

$$\begin{aligned}\mathbf{R}(\tilde{p}, \vec{A}, \vec{B}, \vec{C}) &= \mathbf{R}(p_\mu \tilde{\omega}^\mu, A^\nu \vec{e}_\nu, B^\rho \vec{e}_\rho, C^\sigma \vec{e}_\sigma) \\ &= p_\mu A^\nu B^\rho C^\sigma \mathbf{R}(\tilde{\omega}^u, \vec{e}_\nu, \vec{e}_\rho, \vec{e}_\sigma) \\ &= p_\mu A^\nu B^\rho C^\sigma R^\mu{}_{\nu\rho\sigma}.\end{aligned}$$

An instance of this application was found in the Riemann curvature tensor.

8.3 Exterior Products and Differential Forms

At the outset we can assure our readers that we shall not do away with tensors by introducing differential forms. Tensors are here to stay; in a great many situations, particularly those dealing with symmetries, tensor methods are very natural and effective. However, in many other situations the use of the exterior calculus...leads to decisive results in a way which is very difficult with tensors alone. Sometimes a combination of techniques is in order. –Harley Flanders

In this section we introduce a new kind of vector product, denoted with a wedge and thus called informally the "wedge product." Its formal name is the "exterior product." It generates a new kind of algebra that will also allow a new kind of derivative to be defined, the "exterior derivative." In terms of the exterior derivative, vector identities such as $\boldsymbol{\nabla}\cdot(\boldsymbol{\nabla}\times\mathbf{A}) = 0$ and $\boldsymbol{\nabla}\times(\boldsymbol{\nabla}\phi) = \mathbf{0}$ are unified, as are the integral theorems of Stokes and Gauss. We will focus on the exterior product in this section, turn to the exterior derivative in the next two sections, and conclude with integrals of exterior derivatives in Section 8.6.

The Exterior Product

The notion of the exterior product builds on a familiar player in three-dimensional Euclidean space: the antisymmetric cross product between vectors, $\mathbf{A}\times\mathbf{B} = -\mathbf{B}\times\mathbf{A}$. Two vectors define a plane, but to interpret $\mathbf{A}\times\mathbf{B}$ as a vector perpendicular to that plane, the plane had to be embedded in three-dimensional space. The exterior product, denoted with a wedge $\mathbf{A}\wedge\mathbf{B}$, does away with the necessity for the embedding space.

The wedge product applies to any two vectors $\vec{A}$ and $\vec{B}$ in a space of n dimensions. By definition the wedge product, also called a "bivector," is antisymmetric:

$$\vec{A}\wedge\vec{B} = -\vec{B}\wedge\vec{A}. \tag{8.41}$$

The magnitude $|\vec{A}\wedge\vec{B}|$ of the bivector is the area of the parallelogram defined by the two vectors, although the bivector need not be in the shape of a parallelogram. For example, a circle of radius R has the same area as a square whose sides have length $\sqrt{\pi}R$. The direction of the bivector is equivalent to the right-hand rule, but a normal vector $\hat{\mathbf{n}}$ is not part of a bivector's definition. (By way of analogy, a family sitting down to dinner around the table can define a direction for the table's surface without requiring a normal vector: agree that passing dishes to the right gives the surface a postive sign, and passing to the left gives it a negative sign.)

Let $\mathcal{V}$ be a vector space of n dimensions over the field of real numbers (see Appendix C). In other words, all elements of $\mathcal{V}$ have n components A^μ, each one a real number. For any three vectors $\vec{A}, \vec{B}, \vec{C}$ belonging to $\mathcal{V}$ and for any real numbers α and β, the algebra of the exterior product is defined by its anticommutativity, its associativity, and its being distributive over addition:

$$\begin{aligned}
\vec{A}\wedge\vec{B} &= -\vec{B}\wedge\vec{A} \\
\vec{A}\wedge(\vec{B}\wedge\vec{C}) &= (\vec{A}\wedge\vec{B})\wedge\vec{C} \\
(\alpha\vec{A}+\beta\vec{B})\wedge\vec{C} &= \alpha(\vec{A}\wedge\vec{C})+\beta(\vec{B}\wedge\vec{C}) \\
\vec{C}\wedge(\alpha\vec{A}+\beta\vec{B}) &= \alpha(\vec{C}\wedge\vec{A})+\beta(\vec{C}\wedge\vec{A}).
\end{aligned}$$

Notice that because of its antisymmetry, the wedge product of a vector with itself vanishes:

$$\vec{A}\wedge\vec{A}=0. \tag{8.42}$$

Therefore, any two colinear vectors have null exterior product.

When $\vec{A}$ and $\vec{B}$ are written in terms of basis vectors, their exterior product becomes a superposition of exterior products of the basis vectors:

$$\begin{aligned}
\vec{A}\wedge\vec{B} &= (A^\mu\vec{e}_\mu)\wedge(B^\nu\vec{e}_\nu) \\
&= A^\mu B^\nu(\vec{e}_\mu\wedge\vec{e}_\nu).
\end{aligned}$$

Since $\vec{e}_\mu\wedge\vec{e}_\nu=-\vec{e}_\nu\wedge\vec{e}_\mu$, we may write the superposition over the $\vec{e}_\mu\wedge\vec{e}_\nu$ with antisymmetric tensor coefficients and μ,ν in numerical order, so that

$$\vec{A}\wedge\vec{B}=\sum_{\mu<\nu}(A^\mu B^\nu-A^\nu B^\mu)(\vec{e}_\mu\wedge\vec{e}_\nu). \tag{8.43}$$

Let us work out an example in detail, because the anticommutativity of the wedge product lies at the heart of the matter. Let $\vec{A}=a\vec{e}_1+b\vec{e}_2+c\vec{e}_3$ and $\vec{B}=u\vec{e}_1+v\vec{e}_2+w\vec{e}_3$. Then

$$\begin{aligned}
\vec{A}\wedge\vec{B} &= (a\vec{e}_1+b\vec{e}_2+c\vec{e}_3)\wedge(u\vec{e}_1+v\vec{e}_2+w\vec{e}_3) \\
\\
&= au(\vec{e}_1\wedge\vec{e}_1)+av(\vec{e}_1\wedge\vec{e}_2)+aw(\vec{e}_1\wedge\vec{e}_3) \\
&+ bu(\vec{e}_2\wedge\vec{e}_1)+bv(\vec{e}_2\wedge\vec{e}_2)+bw(\vec{e}_2\wedge\vec{e}_3) \\
&+ cu(\vec{e}_3\wedge\vec{e}_1)+cv(\vec{e}_3\wedge\vec{e}_2)+cw(\vec{e}_3\wedge\vec{e}_3) \\
\\
&= av(\vec{e}_1\wedge\vec{e}_2)+bu(\vec{e}_2\wedge\vec{e}_1) \\
&+ aw(\vec{e}_1\wedge\vec{e}_3)+cu(\vec{e}_3\wedge\vec{e}_1) \\
&+ bw(\vec{e}_2\wedge\vec{e}_3)+cv(\vec{e}_3\wedge\vec{e}_2) \\
\\
&= (av-bu)(\vec{e}_1\wedge\vec{e}_2)+(aw-cu)(\vec{e}_1\wedge\vec{e}_3)+(bw-cv)(\vec{e}_2\wedge\vec{e}_3).
\end{aligned}$$

Notice that the coefficients of the bivectors (with ordered indices) are the components of antisymmetric second-order tensors.

An alternative way of writing $\vec{A} \wedge \vec{B}$ makes use of a cyclic property. Using the antisymmetry of the wedge product, we may also write

$$\begin{aligned} \vec{A} \wedge \vec{B} &= (av - bu)(\vec{e}_1 \wedge \vec{e}_2) + (cu - aw)(\vec{e}_3 \wedge \vec{e}_1) + (bw - cv)(\vec{e}_2 \wedge \vec{e}_3) \\ &\equiv C^3(\vec{e}_1 \wedge \vec{e}_2) + C^2(\vec{e}_3 \wedge \vec{e}_1) + C^1(\vec{e}_2 \wedge \vec{e}_3) \\ &= \sum_{ijk\ cyclic} C^i(\vec{e}_j \wedge \vec{e}_k) \end{aligned}$$

where $C^i \equiv C^{jk} = -C^{kj}$ is the coefficient of $\vec{e}_j \wedge \vec{e}_k$ (notice that ijk are in cyclic order).

There is yet a third way to compress the notation for all the terms in $\vec{A} \wedge \vec{B}$. Suppose we write down all the terms in the sum $C^{jk}(\vec{e}_j \wedge \vec{e}_k)$. Using only the antisymmetry of the wedge product, the nonzero terms would be

$$(C^{12} - C^{21})(\vec{e}_1 \wedge \vec{e}_2) + (C^{13} - C^{31})(\vec{e}_1 \wedge \vec{e}_3) + (C^{23} - C^{32})(\vec{e}_2 \wedge \vec{e}_3). \quad (8.44)$$

But with the C^{jk} also being antisymmetric, this becomes

$$2C^{12}(\vec{e}_1 \wedge \vec{e}_2) + 2C^{13}(\vec{e}_1 \wedge \vec{e}_3) + 2C^{23}(\vec{e}_2 \wedge \vec{e}_3). \quad (8.45)$$

Thus, if we write the wedge product of two vectors in terms of basis vectors without any notes made about cyclical or numerical ordering, a compensating factor of $\frac{1}{2}$ must be included:

$$\vec{A} \wedge \vec{B} = \frac{1}{2} C^{ij}(\vec{e}_i \wedge \vec{e}_j). \quad (8.46)$$

With bivectors in hand, it is irresistible to define a trivector, $\vec{A} \wedge \vec{B} \wedge \vec{C}$, and try to interpret it. The interpretation that works is that of a directed three-dimensional volume. The "directed" part distinguishes right-handed coordinate axes from left-handed ones (see Hestenes, p. 20). With the wedge product one can go on multiplying a trivector with another vector, and so on (with $\frac{1}{r!}$ for an r-vector written in terms of basis vectors, if cyclic or numerical order in the indices is not specified). In a space of n dimensions the series of multivectors ends with a directed volume in n dimensions, a so-called n-vector $\vec{A}_1 \wedge \vec{A}_2 \wedge \ldots \vec{A}_n$ (these subscripts distinguish the vectors; they do not signify basis vectors). The series stops here because in a wedge product of $(n+1)$ vectors in n dimensions, at least two vectors will be colinear and give zero.

Formally, r-vectors for $r = 0, 1, 2, \ldots, n$ can be defined by successive applications of the exterior product, to create the following spaces denoted $\bigwedge^r \mathcal{V}$ (see Flanders, Ch. 2): $\bigwedge^0 \mathcal{V}$, the set of real numbers; the vector space $\mathcal{V}$ itself, $\bigwedge^1 \mathcal{V} = \mathcal{V}$; the set $\bigwedge^2 \mathcal{V}$ of bivectors, containing elements of the form

$$\sum_{\mu<\nu} A^{\mu\nu}(\vec{e}_\mu \wedge \vec{e}_\nu), \quad (8.47)$$

where $A^{\mu\nu} = -A^{\nu\mu}$; and so on. In general, an r-vector is an object from the set $\bigwedge^r \mathcal{V}$ that has the structure

$$\sum A^{\mu_1\mu_2\ldots\mu_r}(\vec{e}_{\mu_1} \wedge \vec{e}_{\mu_2} \wedge \ldots \vec{e}_{\mu_r}), \quad (8.48)$$

where $\mu_1 < \mu_2 < \ldots < \mu_r$, and where $A^{\mu_1\mu_2\ldots\mu_r}$ for $r > 1$ is the component of an order-r tensor that is antisymmetric under the exchange of any two indices. For example, if the μ_k denote the labels 1, 2, 3, 4 in a four-dimensional space, then the sum above for a 3-form includes permutations of the orderings (123), (124), (134), (234) for the wedge products of differentials. Thus, an element of $\bigwedge^3 \mathcal{V}$ for $\mathcal{V}$ having four dimensions would be of the form

$$A^{123}(\vec{e}_1\wedge\vec{e}_2\wedge\vec{e}_3)+A^{124}(\vec{e}_1\wedge\vec{e}_2\wedge\vec{e}_4)+A^{134}(\vec{e}_1\wedge\vec{e}_3\wedge\vec{e}_4)+A^{234}(\vec{e}_2\wedge\vec{e}_3\wedge\vec{e}_4), \quad (8.49)$$

where each A^{ijk} is an algebraic sum of coefficients, antisymmetric under the exchange of any two indices.

Finally, we note that $\bigwedge^r \mathcal{V} = 0$, the null set, if $r > n$ because at least two of the vectors in the exterior product would be identical.

Multivector Algebra

Now an algebra of r-vectors can be defined. If E and F are r-vectors in n-dimensional space, and α is a scalar, then $E + F$ and αE are also r-vectors.

If E is a p-vector and F a q-vector, then $E \wedge F$ is a $(p+q)$-vector, provided that $p+q \leq n$. Notice something interesting about symmetry and antisymmetry: if p and q are *both odd*, then $E\wedge F = -F\wedge E$, as holds for $\vec{A}\wedge\vec{B} = -\vec{B}\wedge\vec{A}$. But if p and q are both even, or if one is odd and the other even, then $A\wedge B = +B\wedge A$ (see Ex. 8.9). In symbols, $E \wedge F = (-1)^{pq} F \wedge E$.

Having defined multiplication of multivectors by the exterior product, an inner product can also be defined for multivectors if the underlying vector space $\mathcal{V}$ has an inner product $\vec{A}\cdot\vec{B}$. Consider a p-vector L,

$$L = \vec{a}_1 \wedge \vec{a}_2 \wedge \ldots \wedge \vec{a}_p, \quad (8.50)$$

and a q-vector M,

$$M = \vec{b}_1 \wedge \vec{b}_2 \wedge \ldots \wedge \vec{b}_q, \quad (8.51)$$

where again these subscripts on $\vec{a}_i$ and $\vec{b}_i$ are merely labels to distinguish various vectors, and do not denote basis vectors. The inner product $L\cdot M$, also denoted $\langle L|M\rangle$, is defined as a determinant of inner products of vectors,

$$\langle L|M\rangle = det(\vec{a}_i \cdot \vec{b}_j) = \langle M|L\rangle, \quad (8.52)$$

where $i = 1, 2, ..., p$ and $j = 1, 2, ...q$.

Differential Forms

Now we are in a position to define *differential* forms. The move from an r-vector to a differential r-form proceeds by rescaling basis vectors with coordinate differentials. These rescaled basis vectors then get multiplied together with the exterior product. In Euclidean space, the rescaled unit vector $\hat{\mathbf{i}}dx \equiv \vec{dx}$

describes a displacement vector parallel to the x-direction. For any coordinate displacement, under any basis, and in any number of dimensions (resorting to tangent spaces for curved manifolds), a basis vector used in differential forms may be articulated as follows: With no sum over μ, let

$$dx^{\mu}\vec{e}_{\mu} \equiv \vec{dx}^{\mu}. \tag{8.53}$$

Analogous to how the component of a tensor of order N was the product of N coordinate displacements, here the differential r-forms are defined as r-vectors made from wedge products of the basis vectors $\vec{dx}^{\mu}$.

A 0-form is a scalar, having no differential basis vectors. A differential 1-form ω (no tilde used here for *differential* forms, although some authors use tildes) has the structure, in any manifold (summing over μ),

$$\omega = A_{\mu}\vec{dx}^{\mu}. \tag{8.54}$$

For example, in Euclidean 3-space,

$$\omega = A_x\vec{dx} + A_y\vec{dy} + A_z\vec{dz}. \tag{8.55}$$

This looks odd; one wants to call the left-hand side $d\omega$ instead of just ω. But the ω as written is the notation of differential forms. As Charles Dickens wrote in another context, "But the wisdom of our ancestors is in the simile; and my unhallowed hands shall not disturb it, or the Country's done for." More will be said about this strange notation when we study integrals of differential forms.

A differential 2-form η looks like this in Euclidean 3-space (written here with the vectors and their coefficients in cyclic order):

$$\eta = C_x(\vec{dy} \wedge \vec{dz}) + C_y(\vec{dz} \wedge \vec{dx}) + C_z(\vec{dx} \wedge \vec{dy}). \tag{8.56}$$

The differential 2-form is the first r-form that can exhibit antisymmetry. Since the C_i in η are to be antisymmetric, we may rename them $C_i \equiv C_{jk} = -C_{kj}$ with ijk in cyclic order, so that η may also be written

$$\eta = C_{yz}(\vec{dy} \wedge \vec{dz}) + C_{zx}(\vec{dz} \wedge \vec{dx}) + C_{xy}(\vec{dx} \wedge \vec{dy}). \tag{8.57}$$

Differential forms of degree 3 or higher (up to n in a space of n dimensions) may be similarly defined.

Now that we have scalars, 1-forms, 2-forms, and so forth, the space of differential multivectors can be defined. Let R denote an ordered set of indices, $R = \{\mu_1, \mu_2, \cdots \mu_r\}$, where $1 \leq \mu_1 < \mu_2 \cdots, \leq \mu_r$. Let $\vec{d^R x}$ denote the ordered product $\vec{dx}^{\mu_1} \wedge \cdots \wedge \vec{dx}^{\mu_r}$, and let A_R be their antisymmetrized (for $r \geq 2$) coefficients (include a factor of $\frac{1}{r!}$ if the μ_i are not in cyclic or numerical order, to avoid double-counting). Consider the sum of multivectors

$$\begin{aligned}\Omega &= \sum_R A_R \vec{d^R x} \\ &= \phi + A_{\mu}\vec{dx}^{\mu} + \frac{1}{2!}C_{\mu\nu}(\vec{dx}^{\mu} \wedge \vec{dx}^{\nu}) + \cdots,\end{aligned}$$

where the last term in the series will be a differential n-vector in n-dimensional space. If two multivectors are equal, then their respective scalar, 1-vector, 2-vector,... "components" are separately equal. Such a set of multivectors forms a "vector space" in the sense of abstract algebra (Appendix C). It is closed under multivector addition and scalar multiplication, respects the associative law under multiplication, and commutes under addition. We started out by asking, "What does 'tensor' mean?" Perhaps we should now be asking, "How far can the meaning of 'vector' be taken?"

Differential forms come into their own when they appear in an integral. We will see that in line integrals, in Stokes's theorem, and in Gauss's divergence theorem, the integral over an r-form gives an $(r-1)$-form (for $r \geq 1$) evaluated on the boundary of the region of integration. But before we learned how to integrate, we had to learn about derivatives. Likewise, before we consider the integral of an r-form, we must discuss its inverse, the "exterior derivative."

8.4 The Exterior Derivative

The exterior derivative of a differential r-form T produces a differential $(r+1)$-form, which carries various notations:

$$\mathbf{d}T \equiv \frac{\partial}{\partial x} \wedge T \equiv \partial \wedge T. \tag{8.58}$$

It is defined as follows (sum over μ):

$$\partial \wedge T \equiv d\vec{x}^{\mu} \wedge (\partial_{\mu} T) \tag{8.59}$$

(notice that a basis vector $\vec{dt} \equiv \vec{dx}^{0}$ is needed for the time direction in space-time).

Let us work our way through some exterior derivatives in three-dimensional space, showing that the exterior derivative of a differential r-form gives a differential $(r+1)$-form. Let ϕ be a scalar, a differential 0-form. Then

$$\partial \wedge \phi = \vec{dx}^{k} (\partial_k \phi), \tag{8.60}$$

which has the structure $A\vec{dx} + B\vec{dy} + C\vec{dz}$, a differential 1-form, where A, B, and C are derivatives of ϕ with respect to coordinates. This is the differential form version of the familiar $d\phi = (\partial_k \phi) dx^k$ well known from the chain rule.

Let ω be a differential 1-form $\omega = A_x \vec{dx} + A_y \vec{dy} + A_z \vec{dz}$, and take its exterior derivative:

$$\begin{aligned}
\partial \wedge \omega &= (\partial_j A_k)(\vec{dx}^j \wedge \vec{dx}^k) \\
&= (\partial_x A_x)(\vec{dx} \wedge \vec{dx}) + (\partial_x A_y)(\vec{dx} \wedge \vec{dy}) + (\partial_x A_z)(\vec{dx} \wedge \vec{dz}) \\
&+ (\partial_y A_x)(\vec{dy} \wedge \vec{dx}) + (\partial_y A_y)(\vec{dy} \wedge \vec{dy}) + (\partial_y A_z)(\vec{dy} \wedge \vec{dz}) \\
&+ (\partial_z A_x)(\vec{dz} \wedge \vec{dx}) + (\partial_z A_y)(\vec{dz} \wedge \vec{dy}) + (\partial_z A_z)(\vec{dz} \wedge \vec{dz}) \\
&= (\partial_y A_z - \partial_z A_y)(\vec{dy} \wedge \vec{dz}) + (\partial_z A_x - \partial_x A_z)(\vec{dz} \wedge \vec{dx}) \\
&+ (\partial_x A_y - \partial_y A_x)(\vec{dx} \wedge \vec{dy}),
\end{aligned}$$

which has the structure of a differential 2-form $T = N(\vec{dx} \wedge \vec{dy}) + M(\vec{dy} \wedge \vec{dz}) + L(\vec{dz} \wedge \vec{dx})$. We recall that T can be written three equivalent ways (watch signs on the coefficients; C_{ij} in the cyclic ordering may not always have the same sign as C_{ij} in the $i < j$ ordering),

$$T = \sum_{\text{I<J}} C_{ij}(\vec{dx}^i \wedge \vec{dx}^j) = \frac{1}{2} C_{ij}(\vec{dx}^i \wedge \vec{dx}^j) = \sum_{ijk\ cyclic} C_{ij}(\vec{dx}^i \wedge \vec{dx}^j), \tag{8.61}$$

because both the second-order tensor coefficients and the wedge products are antisymmetric. In three dimensions the tensor coefficients are the components of the curl of a vector field $\nabla \times \mathbf{C}$.

The exterior derivative of a 2-form, in three-dimensional Euclidean space, gives a result that contains the familiar Euclidean divergence,

$$\partial \wedge T = (\nabla \cdot \mathbf{C})(\vec{dx} \wedge \vec{dy} \wedge \vec{dz}). \tag{8.62}$$

To sum up so far, given a differential form T of degree r,

$$T = \frac{1}{r!} C_{\mu\nu\ldots}(\vec{dx}^\mu \wedge \vec{dx}^\nu \wedge \cdots), \tag{8.63}$$

its exterior derivative is

$$\partial \wedge T = \frac{1}{r!}(\partial_\eta C_{\mu\nu\ldots})(\vec{dx}^\eta \wedge \vec{dx}^\mu \wedge \vec{dx}^\nu \cdots). \tag{8.64}$$

In most literature on this subject, in such expressions the wedge products between differentials are *understood* and not written explicitly. For instance, in the context of differential forms $dydz = -dzdy$ because $dydz$ means $\vec{dy} \wedge \vec{dz} = -\vec{dz} \wedge \vec{dy}$. Omitting the wedges raises no problem if one remembers that the differentials anticommute. The distinction between $\vec{dx}\vec{dy}\vec{dz} = -\vec{dy}\vec{dx}\vec{dz}$ (with wedges understood) on the one hand and the usual volume element $dxdydz = dydxdz$ (no implicit wedges) on the other hand is that the former is a directed volume that distinguishes between right- and left-handed coordinate systems, whereas the latter is merely a number that measures volume. I will continue making the wedges explicit.

Properties of the Exterior Derivative

Let ω be a differential form of degree r, and let η be a differential form of degree q. From its definition one can easily verify the following properties of the exterior derivative:

$$\partial \wedge (\omega + \eta) = (\partial \wedge \omega) + (\partial \wedge \eta), \tag{8.65}$$

$$\partial \wedge (\omega \wedge \eta) = [(\partial \wedge \omega) \wedge \eta] + [(-1)^r \omega \wedge (\partial \wedge \eta)], \tag{8.66}$$

and an especially significant one called Poincaré's lemma, which says that for any differential form T,

$$\partial \wedge (\partial \wedge T) = 0. \tag{8.67}$$

Poincaré's lemma is so important that we should prove it immediately. Let T be a differential r-form,

$$T = \sum_R T_R \vec{d}^R x. \tag{8.68}$$

The first exterior derivative of T gives

$$\begin{aligned} \partial \wedge T &= \vec{dx}^\mu \wedge \partial_\mu T \\ &= \sum_R (\partial_\mu T_R)(\vec{dx}^\mu \wedge \vec{d}^R x). \end{aligned}$$

Now take the second exterior derivative:

$$\begin{aligned} \partial \wedge (\partial \wedge T) &= \vec{dx}^\nu \partial_\nu \wedge \left[\sum_R (\partial_\mu T_R)(\vec{dx}^\mu \wedge \vec{d}^R x)\right] \\ &= \sum_R \frac{\partial^2 T_R}{\partial x^\mu \partial x^\nu}(\vec{dx}^\nu \wedge \vec{dx}^\mu \wedge \vec{d}^R x). \end{aligned}$$

By virtue of continuity of the coefficients T_R, the partial derivatives commute,

$$\frac{\partial^2 T_R}{\partial x^\nu \partial x^\mu} = \frac{\partial^2 T_R}{\partial x^\mu \partial x^\nu}. \tag{8.69}$$

However,

$$\vec{dx}^\nu \wedge \vec{dx}^\mu = -\vec{dx}^\mu \wedge \vec{dx}^\nu, \tag{8.70}$$

and thus $\partial \wedge (\partial \wedge T) = -\partial \wedge (\partial \wedge T) = 0$, QED.

Applications of $\partial \wedge (\partial \wedge T) = 0$ are immediate and easily demonstrated in three-dimensional Euclidean space (see Exs. 8.1 and 8.2). For instance, if T is a 0-form ϕ, then

$$0 = \partial \wedge (\partial \wedge \phi) = \nabla \times (\nabla \phi). \tag{8.71}$$

If T is a 1-form $\mathbf{v}$, then

$$0 = \partial \wedge (\partial \wedge \mathbf{v}) = \boldsymbol{\nabla} \cdot (\boldsymbol{\nabla} \times \mathbf{v}). \tag{8.72}$$

Conspicuously, the antisymmetric curl appears in both of these vector theorems.

A corollary to Poincaré's lemma immediately follows: if σ is a differential r-form and

$$\partial \wedge \sigma = 0 \tag{8.73}$$

then a differential $(r-1)$-form ω exists such that

$$\sigma = \partial \wedge \omega. \tag{8.74}$$

The corollary seems almost obvious, given Poincaré's lemma, but proving it rigorously takes one into the deep waters of articulating the circumstances under which it can happen. The proof (see Flanders, Ch. 3) shows it to be valid only in domains that are topologically simple, that can be deformed in principle into a point (e.g., a toroid cannot because however you stretch or distort it without tearing, a hole remains).

Let us consider a sample application of differential forms. Whether or not this book is "concise," at least it should include concise applications of the methods it discusses.

8.5 An Application to Physics: Maxwell's Equations

Physics in Flat Spacetime:
Wherein the reader meets an old friend, Special Relativity, outfitted in a new, mod attire, and becomes more intimately acquainted with her charms.
–Misner, Thorne, and Wheeler, about to introduce special relativity in the language of differential forms

Consider the electric fields in matter, $\mathbf{E}$ and $\mathbf{D}$, and the magnetic fields in matter, $\mathbf{B}$ and $\mathbf{H}$. The fields $\mathbf{D}$ and $\mathbf{H}$ are due to free charges and currents, and the fields $\mathbf{E}$ and $\mathbf{B}$ are due to *all* sources of fields, including the effects of polarization. The homogeneous Maxwell equations are

$$\boldsymbol{\nabla} \times \mathbf{E} = -\frac{\partial \mathbf{B}}{\partial t} \tag{8.75}$$

and

$$\boldsymbol{\nabla} \cdot \mathbf{B} = 0. \tag{8.76}$$

The inhomogeneous equations, which relate the fields directly to the density ρ of free charges and the density $\mathbf{j}$ of free currents, are

$$\boldsymbol{\nabla} \times \mathbf{H} = \mathbf{j} + \frac{\partial \mathbf{D}}{\partial t} \tag{8.77}$$

and

$$\nabla \cdot \mathbf{D} = \rho. \tag{8.78}$$

Let us examine unified treatments of Maxwell's electrodynamics, first recalling the language of tensors, and then after that using the language of differential forms.

In terms of tensors, we define the 4-vectors $A^\mu = (\phi, \mathbf{A})$ for the potentials and $j^\mu = (\rho, \mathbf{j})$ for the sources, along with their duals $A_\mu = (\phi, -\mathbf{A})$ and $j_\mu = (\rho, -\mathbf{j})$, the gradient $\partial_\mu = (\partial_t, \nabla)$, and its dual $\partial^\mu = (\partial_t, -\nabla)$. From these comes the Faraday tensor, $F^{\mu\nu} = \partial^\mu A^\nu - \partial^\nu A^\mu$. The inhomogeneous Maxwell equations (absorbing constants) then take the form

$$\partial_\mu F^{\mu\nu} = j^\nu. \tag{8.79}$$

The homogeneous Maxwell equations may be written

$$\sum_{\mu\nu\tau \text{ cyclic}} \partial_\mu F_{\nu\tau} = 0. \tag{8.80}$$

The tensor formalism is elegant and useful. But here is something else elegant and useful: Armed with the exterior product, introduce a time-like basis vector $\vec{dt} = dt\vec{e}_0$, so that exterior products for *spacetime* can be defined. In particular, define a 2-form α,

$$\begin{aligned}
\alpha &= E_k(\vec{dx}^k \wedge \vec{dt}) + \sum_{ijk \text{ cyclic}} B_i(\vec{dx}^j \wedge \vec{dx}^k) \\
&= [E_x\vec{dx} + E_y\vec{dy} + E_z\vec{dz}] \wedge \vec{dt} + B_x(\vec{dy} \wedge \vec{dz}) + B_y(\vec{dz} \wedge \vec{dx}) + B_z(\vec{dx} \wedge \vec{dy}),
\end{aligned}$$

and another 2-form β,

$$\beta = -(H_x\vec{dx} + H_y\vec{dy} + H_z\vec{dz}) \wedge \vec{dt} + [D_x(\vec{dy} \wedge \vec{dz}) + D_y(\vec{dz} \wedge \vec{dx}) + D_z(\vec{dx} \wedge \vec{dy})]. \tag{8.81}$$

For the charge and current sources of the fields introduce a third differential 2-form γ,

$$\gamma = [j_x(\vec{dy} \wedge \vec{dz}) + j_y(\vec{dz} \wedge \vec{dx}) + j_z(\vec{dx} \wedge \vec{dy})] \wedge \vec{dt} - \rho(\vec{dx} \wedge \vec{dy} \wedge \vec{dz}). \tag{8.82}$$

Now comes the fun part. Evaluate $\partial \wedge \alpha$. Writing the spatial indices with Latin letters in cyclic order, there results

$$\begin{aligned}
\partial \wedge \alpha &= \vec{dx}^\nu \wedge \partial_\nu \alpha \\
&= \vec{dt} \wedge (\partial_t \alpha) + \vec{dx}^i \wedge (\partial_i \alpha) \\
&= \left[(\nabla \times \mathbf{E})_i + \frac{\partial B_i}{\partial t}\right](\vec{dx}^j \wedge \vec{dx}^k \wedge \vec{dt}) + (\nabla \cdot \mathbf{B})(\vec{dx} \wedge \vec{dy} \wedge \vec{dz}).
\end{aligned}$$

By virtue of Maxwell's equations, Eqs. (8.75) and (8.76), Faraday's law and Gauss's law for the magnetic field are concisely subsumed into one equation by setting

$$\partial \wedge \alpha = 0. \tag{8.83}$$

Similar treatment for the inhomogeneous Maxwell equations (Gauss's law for the electric field, and the Ampère-Maxwell law) are left as exercises for you to enjoy (see Exs. 8.4-8.7).

8.6 Integrals of Differential Forms

Since the exterior derivative of a differential r-form produces a differential $(r+1)$-form, we expect the integral of a differential r-form to produce a differential $(r-1)$-form (for $r > 0$). When discussing antiderivatives of exterior derivatives, the notation $\mathbf{d}T$ is more convenient than $\partial \wedge T$, even though their meanings are the same.

We might expect that the integral of $\mathbf{d}\omega$ would simply be ω, but instead in the literature on integrating differential forms we encounter this weird-looking formula:

$$\int_{\mathcal{R}} \mathbf{d}\omega = \int_{bd\ \mathcal{R}} \omega. \tag{8.84}$$

Eq. (8.84) looks unsettling–and can be distracting–because there seems to be a d missing in the integral on the right. This oddness echoes the seemingly "missing d" that was mentioned after Eq. (8.55). Here is a notational threshold to get over! Of course, in the strange notation of $\int_{bd\mathcal{R}} \omega$, the differentials are already built into ω and are joined together by wedge products if ω has degree $r > 1$. Thus, $\int_{bd\mathcal{R}} \omega$ is indeed an integral over r differentials, with the $\vec{dx}^{\mu}$ differentials understood. But there is more to it than implicit d's. Let us look for something analogous in elementary calculus.

In calculus we learned that

$$\int_a^b dx = x|_a^b = x(b) - x(a). \tag{8.85}$$

The term on the right-hand side of Eq. (8.84) means something analogous to the $x|_a^b$ of elementary calculus. Indeed, they would look almost the same if we wrote $x|_a^b$ as $|_a^b x$. In both $x|_a^b$ and $\int_{bd\ \mathcal{R}} \omega$, the result of evaluating the definite integral over a region $\mathcal{R}$ is that the antiderivative gets evaluated on the boundary of $\mathcal{R}$.

In an elementary integral $\int_a^b f(x)dx$, $\mathcal{R}$ is a piece of x-axis from $x = a$ to $x = b$. The boundary of a non-closed surface (e.g., a flux integral) is a closed contour; the boundary of a volume is the surface enclosing the volume. Whatever region $\mathcal{R}$ happens to be, $\int_{bd\mathcal{R}} \omega$ means that ω gets evaluated on the boundary of $\mathcal{R}$.

Let me be more explicit. Consider a line integral, such as work done by a conservative force $\mathbf{F}$ as a particle moves from $\mathbf{r} = \mathbf{a}$ to $\mathbf{r} = \mathbf{b}$, so that

$$\begin{aligned}\int_{\mathbf{a}}^{\mathbf{b}} \mathbf{F} \cdot d\mathbf{r} &= -\int_{\mathbf{a}}^{\mathbf{b}} dU \\ &= -[U(\mathbf{b}) - U(\mathbf{a})].\end{aligned}$$

Also recall Stokes's theorem, where $\mathcal{C}$ is the closed path that forms the boundary of surface $\mathcal{S}$,

$$\int_{\mathcal{S}} (\boldsymbol{\nabla} \times \mathbf{B}) \cdot \hat{\mathbf{n}} da = \oint_{C} \mathbf{B} \cdot d\mathbf{r}, \tag{8.86}$$

and the divergence theorem, where closed surface Γ is the boundary of volume $\mathcal{V}$,

$$\int_{\mathcal{V}} (\boldsymbol{\nabla} \cdot \mathbf{E}) d^3 r = \oint_{\Gamma} \mathbf{E} \cdot \hat{\mathbf{n}} da. \tag{8.87}$$

The line integral, Stokes's theorem, and the divergence theorem are familiar from vector calculus in three-dimensional Euclidean space. However, these three integral theorems are merely special cases of Eq. (8.84). Let us demonstrate that now, starting with path integrals. The beauty of differential forms is that such results can be extended to manifolds of n dimensions and non-Euclidean metrics.

In ordinary calculus, we have the familiar result for a scalar function ϕ,

$$\int_a^b d\phi = \phi|_a^b. \tag{8.88}$$

Let us write its differential form analog. Being a scalar, ϕ is also a differential 0-form. Its exterior derivative is a 1-form,

$$\mathbf{d}\phi = (\partial_\mu \phi) \vec{dx}^{\mu}. \tag{8.89}$$

Now let us undo the derivative and integrate it from $\vec{x} = \vec{a}$ to $\vec{x} = \vec{b}$:

$$\int_{\vec{a}}^{\vec{b}} \mathbf{d}\phi = \phi|_{\vec{a}}^{\vec{b}} \equiv \int_{endpoints} \phi. \tag{8.90}$$

The odd notation of Eq. (8.84) has the meaning of the middle term, and the right-hand term is an instance of $\int_{bd\ \mathcal{R}} \omega$. Now let us carry out the same procedure on higher-degree differential forms.

Consider a differential 1-form ψ:

$$\psi = B_\mu \vec{dx}^{\mu}. \tag{8.91}$$

Evaluate its exterior derivative:

$$\mathbf{d}\psi = (\partial_\nu B_\mu)(\vec{dx}^{\nu} \wedge \vec{dx}^{\mu}). \tag{8.92}$$

Allowing for the antisymmetry of the wedge product, this may be written

$$\mathbf{d}\psi = \sum_{\mu<\nu} (\partial_\mu B_\nu - \partial_\nu B_\nu)(\vec{dx}^{\mu} \wedge \vec{dx}^{\nu}). \tag{8.93}$$

Now undo the derivative by evaluating the integral of $\mathbf{d}\psi$ over a region $\mathcal{R}$. That yields ψ evaluated on the boundary of $\mathcal{R}$:

$$\int_{\mathcal{R}} \mathbf{d}\psi = \int_{bd\ \mathcal{R}} \psi. \tag{8.94}$$

Since $\mathbf{d}\psi$ has two differentials, $\mathcal{R}$ is a surface $\mathcal{S}$. Since ψ has one differential, $bd\ \mathcal{R}$ is a closed contour $\mathcal{C}$. Putting these limits into the integrals, and restoring what $\mathbf{d}\psi$ and ψ are in terms of differentials with coefficients, Eq. (8.94) says

$$\int_{\mathcal{S}} \sum_{\mu<\nu} (\partial_\mu B_\nu - \partial_\nu B_\nu)(\vec{dx}^\mu \wedge \vec{dx}^\nu) = \oint_{\mathcal{C}} B_\mu \vec{dx}^\mu, \tag{8.95}$$

which will be recognized as Stokes's theorem, Eq. (8.86), in the language of differential forms.

Let η be a differential 2-form,

$$\eta = \frac{1}{2} C_{\mu\nu} (\vec{dx}^\mu \wedge \vec{dx}^\nu), \tag{8.96}$$

where $C_{\mu\nu} = -C_{\nu\mu}$. Evaluate its exterior derivative:

$$\mathbf{d}\eta = \frac{1}{2} (\partial_\rho C_{\mu\nu})(\vec{dx}^\rho \wedge \vec{dx}^\mu \wedge \vec{dx}^\nu). \tag{8.97}$$

Allowing for the antisymmetry of both the wedge product and the second-order tensor coefficients, this becomes $(\partial_1 C_{23} + \partial_2 C_{31} + \partial_3 C_{12})(\vec{dx} \wedge \vec{dy} \wedge \vec{dz})$. Finally, denoting $C_\rho \equiv C_{\mu\nu}$ with ρ, μ, ν in cyclic order, we obtain

$$\mathbf{d}\eta = (\partial_\rho C_\rho)(\vec{dx} \wedge \vec{dy} \wedge \vec{dz}). \tag{8.98}$$

Now let us reverse what we just did and integrate Eq. (8.98),

$$\int_{\mathcal{R}} \mathbf{d}\eta = \int_{bd\ \mathcal{R}} \eta, \tag{8.99}$$

restore what $\mathbf{d}\eta$ and η are in terms of differential forms, and write volume $\mathcal{V}$ for $\mathcal{R}$ and surface $\mathcal{S}$ for its boundary:

$$\int_{\mathcal{V}} (\partial_\rho C_\rho)(\vec{dx} \wedge \vec{dy} \wedge \vec{dz}) = \oint_{\mathcal{S}} \sum_{\mu<\nu} C_{\mu\nu} (\vec{dx}^\mu \wedge \vec{dx}^\nu). \tag{8.100}$$

In three-dimensional Euclidean space, this is the differential form rendition of the divergence theorem, Eq. (8.87).

Thus, we have seen, for any differential form ω, that Poincaré's lemma,

$$\mathbf{d}(\mathbf{d}\omega) = 0, \tag{8.101}$$

is equivalent to the vector identities in Euclidean space

$$\begin{aligned} \boldsymbol{\nabla} \cdot (\boldsymbol{\nabla} \times \mathbf{A}) &= 0 \\ \boldsymbol{\nabla} \times (\boldsymbol{\nabla} \phi) &= \mathbf{0} \end{aligned}$$

for any vector $\mathbf{A}$ and any scalar ϕ, and that the integral

$$\int_{\mathcal{R}} \mathbf{d}\omega = \int_{bd\ \mathcal{R}} \omega \tag{8.102}$$

contains line integrals, Stokes's theorem, and Gauss's divergence theorem as special cases.

Having announced that "tensors are here to stay," Flanders also comments (p. 4) that "exterior calculus is here to stay, that it will gradually replace tensor methods in numerous situations *where it is the more natural tool* [emphasis added], that it will find more and more applications because of its inner simplicity,...and because it simply is there whenever integrals occur....Physicists are beginning to realize its usefulness."

Postscript

The fundamental principles of physics must transcend this or that coordinate system and therefore be amendable to being written in *generalized* coordinates. For any equation to be written with the convenience of coordinates, while also transcending them, it must be written as tensors. Tensors transform the same as the coordinates themselves, and therefore move smoothly between coordinate systems. That is why tensors are formally defined by how they behave under coordinate transformations!

Tensors and their friends, such as the affine connection and differential forms, can get complicated when working out their details. But the underlying ideas are simple and elegant, beautiful products of the human imagination. Their applications are informed by the real world that we attempt to model with these tools.

Physics is the art of creating, testing, and improving a network of concepts in terms of which nature becomes comprehensible. Because nature is subtle but quantitative, we need robust mathematical tools with which to engage her.

8.7 Discussion Questions and Exercises

Discussion Questions

Q8.1 A tensor can be defined as a mapping from a set of vectors to the real numbers. Is it possible to map a tensor to another tensor? Consider, for example, the prospect of leaving one or more input slots empty when inserting vectors into a tensor of order greater than 2, such as $\mathbf{T}(\ ,\ ,\vec{A})$. What sort of mathematical object is this? What sort of object results from inserting two vectors into two input slots of the Riemann tensor, $\mathbf{R}(\ ,\ ,\vec{A},\vec{B})$, leaving two unfilled slots?

Q8.2 Given the definition of a bivector $\vec{A} \wedge \vec{B}$, consider yet another kind of vector product, simply called "the vector product" denoted by juxtaposition

$\vec{A}\vec{B}$ (not a dyad). This vector product is defined by giving it a scalar and a bivector part:

$$\vec{A}\vec{B} \equiv (\vec{A} \cdot \vec{B}) + (\vec{A} \wedge \vec{B}) \tag{8.103}$$

(see Hestenes). Multiplication among multivectors is defined by extending this idea. If Λ_r is an r-vector, so that $\Lambda_r = \vec{A}_1 \wedge \vec{A}_2 \wedge \ldots \vec{A}_r$, where $r \leq n$ in n-dimensional space (the subscripts label distinct vectors and are not basis vectors) and $\vec{A}$ is another vector, then

$$\vec{A}\Lambda_r \equiv (\vec{A} \cdot \Lambda_r) + (\vec{A} \wedge \Lambda_r), \tag{8.104}$$

which produces an $(r-1)$-vector $\vec{A} \cdot \Lambda_r$ and an $(r+1)$-vector $\Lambda_{r+1} = \vec{A} \wedge \Lambda_r$ (recall that the wedge product of a vector with an n-vector is defined to be zero). Each multivector has a sense of direction, such as a right-hand rule that orients the surface of a bivector, and the distinction between left-and right-handed coordinate systems in trivectors.

A contraction of bivectors follows if an inner product exists between vectors, for example,

$$(\vec{a} \wedge \vec{b}) \cdot (\vec{u} \wedge \vec{v}) \equiv \begin{vmatrix} \vec{a} \cdot \vec{u} & \vec{a} \cdot \vec{v} \\ \vec{b} \cdot \vec{u} & \vec{b} \cdot \vec{v} \end{vmatrix}. \tag{8.105}$$

Another relation is

$$\Lambda_r \wedge \Lambda_s = (-1)^{rs} \Lambda_s \wedge \Lambda_r. \tag{8.106}$$

How is this "geometric algebra" similar to the exterior product discussed in this chapter? The generalized multivector, with scalar, vector, bivector, trivector... "components," is said to form a "tensor basis." Is this name justified?

Motivation for multivector approaches comes with the ability of these algebras to treat vectors and tensors, as well as spinors and hypercomplex numbers (such as the Pauli spinors and Dirac γ-matrices for relativistic electron theory), in a unified formalism (see Hestenes, Yaglom).

Q8.3 For another example where skew-symmetry plays a revealing role, and thereby motivates the wedge product and its use in differential forms, consider an ordinary double integral

$$\int A(x,y)dxdy \tag{8.107}$$

and carry out a change of variable,

$$\begin{aligned} x &= x(u,v) \\ y &= y(u,v), \end{aligned}$$

so that

$$\begin{aligned} dx &= \frac{\partial x}{\partial u}du + \frac{\partial x}{\partial v}dv \\ dy &= \frac{\partial y}{\partial u}du + \frac{\partial y}{\partial v}dv. \end{aligned}$$

Upon substituting these expressions for dx and dy and gathering terms, the integral takes the form

$$\begin{aligned}\int A(x,y)dxdy &= \int A(x(u,v),y(u,v)) \\ &\times \left[\frac{\partial x}{\partial u}\frac{\partial y}{\partial u}dudu + \frac{\partial x}{\partial u}\frac{\partial y}{\partial v}dudv + \frac{\partial x}{\partial v}\frac{\partial y}{\partial u}dvdu + \frac{\partial x}{\partial v}\frac{\partial y}{\partial v}dvdv\right].\end{aligned} \tag{8.108}$$

If we could set

$$dudu = 0 = dvdv \tag{8.109}$$

and

$$dudv = -dvdu, \tag{8.110}$$

then the integral would become

$$\int A(x,y)dxdy = \int A(x(u,v),y(u,v))Jdudv, \tag{8.111}$$

where J is the determinant of the transformation coefficients:

$$J \equiv \begin{vmatrix} \frac{\partial x}{\partial u} & \frac{\partial x}{\partial v} \\ \frac{\partial y}{\partial u} & \frac{\partial y}{\partial v} \end{vmatrix}. \tag{8.112}$$

Show how this procedure could be formalized with the exterior product.

Q8.4 The de Broglie hypothesis of quantum mechanics postulates that corresponding to the motion of a free particle of momentum $\mathbf{p}$, there is a harmonic wave of wavelength λ, where $p = h/\lambda$ and h is Planck's constant. Discuss how this hypothesis could be eloquently expressed in the language of a 1-form. See Misner, Thorne, and Wheeler, p. 53.

Q8.5 Let U denote the xy plane and V the real number line visualized along the z-axis. Is three-dimensional Euclidean space equivalent to $U \otimes V$?

Exercises

8.1 Consider a scalar ϕ, a 0-form. (a) Show that $\partial \wedge \phi$ is a 1-form.
(b) Show that $\partial \wedge (\partial \wedge \phi) = 0$ and demonstrate its equivalence to $\boldsymbol{\nabla} \times (\boldsymbol{\nabla}\phi) = \mathbf{0}$.

8.2 Let ω be a 1-form,

$$\omega = A\vec{dx} + B\vec{dy} + C\vec{dz}. \tag{8.113}$$

(a) Show that $\partial \wedge \omega$ is a 2-form.
(b) Show that $\partial \wedge (\partial \wedge \omega) = 0$ and is equivalent to $\boldsymbol{\nabla} \cdot (\boldsymbol{\nabla} \times \mathbf{A}) = 0$.

8.3 Consider a 2-form T,

$$T = \frac{1}{2} C_{\rho\sigma} (\vec{dx}^{\rho} \wedge \vec{dx}^{\sigma}) \tag{8.114}$$

where $C_{\rho\sigma} = -C_{\sigma\rho}$. Show that $\partial \wedge T$ is a 3-form that contains the divergence of a vector $\mathbf{C}$.

Exs. 8.4 - 8.8 are adapted from Flanders:

8.4 Fill in the steps leading to $\partial \wedge \alpha$ for the α of Section 8.5 which subsumes Faraday's law and Gauss's law for $\mathbf{B}$, viz., Eqs. (8.75) and (8.76).

8.5 Consider the differential 2-form β of Eq. (8.81) and the differential 3-form γ of Eq. (8.82). Show how they can be used to express the inhomogeneous Maxwell equations, Eqs. (8.77) and (8.78).

8.6 Compute $\partial \wedge \gamma$ for the γ of Eq. (8.82) and show how it is related to the local conservation of electric charge,

$$\nabla \cdot \mathbf{j} + \frac{\partial \rho}{\partial t} = 0. \tag{8.115}$$

8.7 Consider the electromagnetic potential 4-vector $A_\mu = (\phi, -\mathbf{A})$. Construct the differential form

$$\lambda = A_\mu \vec{dx}^{\mu} = \phi \vec{dt} + A^i \vec{dx}^{i}. \tag{8.116}$$

Show that $\partial \wedge \lambda = -\alpha$ is equivalent to $\mathbf{E} = -\boldsymbol{\nabla}\phi - \partial_t \mathbf{A}$ and $\mathbf{B} = \boldsymbol{\nabla} \times \mathbf{A}$, where α is given by Eq. (8.81).

8.8 Consider the differential forms made from the electromagnetic fields $\mathbf{E}$, $\mathbf{D}$, $\mathbf{B}$, and $\mathbf{H}$, and the current density $\mathbf{j}$:

$$\begin{aligned}
\alpha &= E_i \vec{dx}^{i} \\
\beta &= \sum_{ijk\ cyclic} B_i (\vec{dx}^{j} \wedge \vec{dx}^{k}) \\
\gamma &= H_i \vec{dx}^{i} \\
\phi &= \sum_{ijk\ cyclic} D_i (\vec{dx}^{j} \wedge \vec{dx}^{k}) \\
\zeta &= \sum_{ijk\ cyclic} j_i (\vec{dx}^{j} \wedge \vec{dx}^{k}).
\end{aligned}$$

(a) Show that

$$\alpha \wedge \beta = \sum_{ijk\ cyclic} S_i (\vec{dx}^{j} \wedge \vec{dx}^{k}), \tag{8.117}$$

where S_i denotes a component of Poynting's vector $\mathbf{S} = \mathbf{E} \times \mathbf{H}$.
(b) Define $\partial' \wedge T$ as the exterior derivative over spatial components only. Show that Maxwell's equations may be written

$$\begin{aligned} \partial' \wedge \phi &= \rho(\vec{dx} \wedge \vec{dy} \wedge \vec{dz}) \\ \partial' \beta &= 0 \\ \partial' \wedge \gamma &= \zeta + \partial_t \phi \\ \partial' \wedge \alpha &= -\partial_t \beta, \end{aligned}$$

where ρ denotes the electric charge density.
(c) Recall Poyning's theorem, written conventionally as

$$\nabla \cdot \mathbf{S} + \frac{\partial \eta}{\partial t} = -\mathbf{j} \cdot \mathbf{E}, \tag{8.118}$$

where η is the electromagnetic energy density, $\eta = \frac{1}{2}\mathbf{E} \cdot \mathbf{D} + \frac{1}{2}\mathbf{B} \cdot \mathbf{H}$. Show that Poynting's theorem may be written in terms of differential forms by applying the product rule for the exterior derivative, Eq. (8.66), applied to $\partial'(\alpha \wedge \gamma)$.

8.9 Let E be a p-vector and F be a q-vector. Show that $E \wedge F = (-1)^{pq} F \wedge E$, in particular:
(a) if p and q are *both odd*, then $E \wedge F = -F \wedge E$;
(b) if p and q are both even, then $E \wedge F = F \wedge E$; and
(c) if p is odd and q is even, then $E \wedge F = F \wedge E$.

Appendix A

Common Coordinate Systems

Over a dozen coordinate systems exist for mapping locations in three-dimensional Euclidean space. The systems most commonly used are rectangular (Cartesian), spherical, and cylindrical coordinates. The cylindrical coordinates (ρ, θ, z) and the spherical coordinates (r, θ, φ) are related to the Cartesian coordinates (x, y, z) as follows (see Fig. A.1).

Cylindrical/Rectangular

$$\begin{aligned} x &= \rho \cos\theta \\ y &= \rho \sin\theta \\ z &= z; \end{aligned}$$

Spherical/Rectangular

$$\begin{aligned} x &= r \sin\theta \cos\varphi \\ y &= r \sin\theta \sin\varphi \\ z &= r \cos\theta. \end{aligned}$$

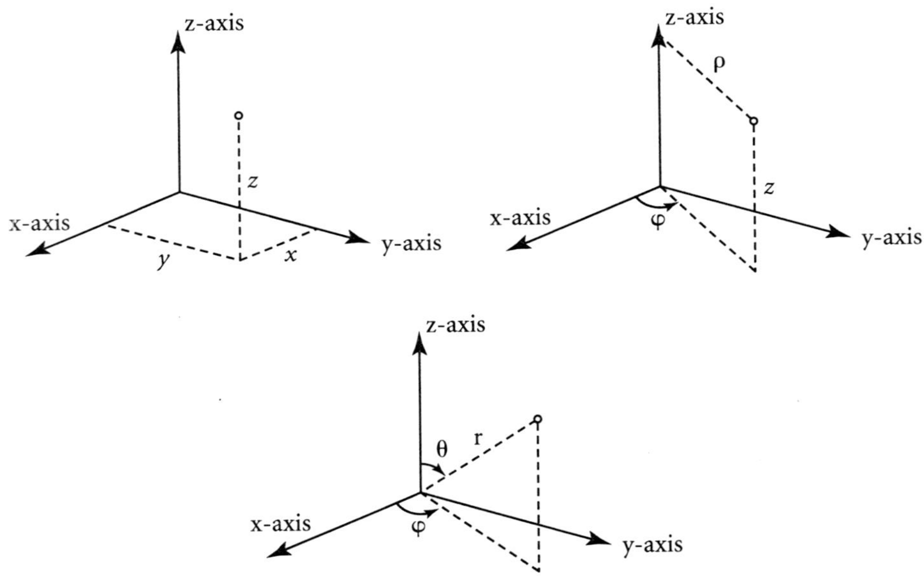

Figure A.1: *Rectangular coordinates (upper left) mapped to cylindrical coordinates (upper right) and spherical coordinates (bottom).*

Appendix B

Theorem of Alternatives

Let M be a square matrix, let $|y\rangle$ be a known vector, and let $|x\rangle$ be an unknown vector. Consider three types of matrix equations to be solved for $|x\rangle$: the inhomogeneous (I), homogeneous (H), and transposed (T) cases,

$$M|x_I\rangle = |y\rangle \tag{B.1}$$

$$M|x_H\rangle = |0\rangle \tag{B.2}$$

$$M^\dagger|x_T\rangle = |0\rangle \tag{B.3}$$

where the dagger denotes the "adjoint," the complex conjugate of the transpose of M, and $|0\rangle$ denotes the null vector.

Because the determinant of M either does or does not vanish, the "theorem of alternatives" (a.k.a. the "invertible matrix theorem") offers the various kinds of solutions that may be found to (I), (H), and (T). The theorem says:

If $|M| \neq 0$, there exists unique solutions as given by

$$\begin{aligned} |x_I\rangle &= M^{-1}|y\rangle \\ |x_H\rangle &= |0\rangle \\ |x_T\rangle &= |0\rangle. \end{aligned}$$

If $|M| = 0$, then there exists nontrival solutions

$$|x_H\rangle \neq |0\rangle \tag{B.4}$$

$$|x_T\rangle \neq |0\rangle \tag{B.5}$$

and there exists solutions $|x_I\rangle$ if and only if

$$\langle x_T|y\rangle = 0 \tag{B.6}$$

for every $|x_T\rangle \neq |0\rangle$.

The part of the theorem we need for the eigenvalue problem is the homogeneous case. For a proof and more discussion of the theorem of alternatives, see textbooks on linear algebra (e.g., Lay, p. 112).

Appendix C

Abstract Vector Spaces

In the context of abstract algebraic systems, a vector space may be described as follows:

Let $\mathcal{V} = \{\alpha, \beta, \ldots\}$ denote a set of elements called "vectors" on which has been defined a binary operation called "vector addition." Also the "scalar multiplication" of elements of $\mathcal{V}$ by the elements $(x, y, \ldots)$ from a field $\mathcal{F}$ is well defined, so that $x\alpha$ is an element of $\mathcal{V}$ for any element x of $\mathcal{F}$ and any elements α of $\mathcal{V}$. Then $\mathcal{V}$ is an abstract *linear space*, or *vector space* over $\mathcal{F}$, if and only if the following conditions are satisfied for arbitrary elements α, β of $\mathcal{V}$, and arbitrary elements x, y of $\mathcal{F}$.

Vector addition is commutative,

$$\alpha + \beta = \beta + \alpha. \tag{C.1}$$

Addition also distributes over multiplication by elements of $\mathcal{F}$, so that

$$\begin{aligned} x(\alpha + \beta) &= x\alpha + x\beta \\ (x + y)\alpha &= x\alpha + y\alpha \\ (xy)\alpha &= (x)(y\alpha) \\ 1\alpha &= \alpha. \end{aligned}$$

Bibliography

Abers, E. S., and Lee, B. W., "Gauge Theories," *Phys. Rep.* **96**, 1–141 (1973).

Acheson, D. J., *Elementary Fluid Mechanics*, (Clarendon Press, Oxford, UK, 1990).

Aitchison, I. J. R., and Hey, A. J. G., *Gauge Theories in Particle Physics* (Adam Hilger Ltd., Bristol, UK, 1982).

Berger, Marcel, *A Panoramic View of Riemannian Geometry* (Springer-Verlag, Heidelberg, Germany, 2003).

Bergmann, Peter G., *Introduction to the Theory of Relativity* (Dover, New York, NY, 1976; Prentice-Hall, Englewood Cliffs, NJ, 1942).

Bjorken, James D., and Drell, Sidney D., *Relativistic Quantum Mechanics* (McGraw-Hill, New York, NY, 1964).

Bloch, Felix, "Heisenberg and the Early Days of Quantum Mechanics," *Physics Today*, December 1976, 23-27.

Boas, Mary L., *Mathematical Methods in the Physical Sciences* (Wiley, New York, NY, 1966).

Bradbury, T. C., *Theoretical Mechanics* (John Wiley and Sons, New York, NY, 1968).

Bronowski, Jacob, *Science and Human Values* (Harper and Row, New York, NY, 1956.

Cartan, Élie, *The Theory of Spinors* (Dover Publications, New York, NY, 1966).

Charap, John M., *Covariant Electrodynamics* (Johns Hopkins University Press, Baltimore, MD, 2011).

Corson, Dale R., Lorrain, Paul, and Lorrain, Francois, *Fundamentals of Electromagnetic Phenomena* (W. H. Freeman, New York, NY, 2000).

Courant, R., and Hilbert, David, *Methods of Mathematial Physics, Vol. 1* (Interscience Publishers, New York, NY, 1953).

Dallen, Lucas, and Neuenschwander, D. E., "Noether's Theorem in a Rotating Reference Frame," *Am. J. Phys.* **79** (3), 326-332 (2011).

Dirac, P. A. M., *General Theory of Relativity* (Wiley, New York, NY, 1975).

Dirac, P. A. M., *The Principles of Quantum Mechanics*, 3rd. ed. (Oxford University Press, London, UK, 1947).

Eddington, Arthur, *The Mathematical Theory of Relativity* (Cambridge University Press (US branch), New York, NY, 1965; Cambridge University Press, Cambridge, UK, 1923).

Einstein, Albert, *The Meaning of Relativity* (Princeton University Press, Princeton, NJ, 1922).

Einstein, Albert, Lorentz, H. A., Weyl, H., and Minkowski, H., *The Principle of Relativity: A Collection of Papers on the Special and General Theory of Relativity* (Dover Publications, New York, NY, 1952; Methuen and Co., London, UK, 1932).

Feynman, Richard P., Leighton, Robert B., and Sands, Matthew, *The Feynman Lectures on Physics, Vols. I-III* (Addison-Wesley, Reading, MA, 1964).

Flanders, Harley, *Differential Forms with Applications to the Physical Sciences* (Dover Publications, New York, NY, 1989; original Academic Press, New York, NY, 1963).

French, A. P., *Special Relativity* (W. W. Norton, New York, NY, 1968).

Goldstein, Herbert, *Classical Mechanics* (Addison-Wesley, Reading, MA, 1950).

Griffiths, David J., *Introduction to Electrodynamics*, 3rd ed. (Prentice Hall, Upper Saddle River, NJ, 1999).

Hartle, James B., *Gravity: An Introduction to Einstein's General Relativity* (Addison-Wesley, San Francisco, CA, 2003).

Hestenes, David, *Space-Time Algebra* (Gordon and Breach, New York, NY, 1966).

Hobson, M. P., Efstathiou, G., and Lasenby, A. N.,*General Relativity: An Introduction for Physicists* (Cambridge University Press, Cambridge, UK, 2006).

Jackson, John D. *Classical Electrodynamics* (Wiley, New York, NY, 1975).

Kobe, Donald,"Generalization of Coulomb's Law to Maxwell's Equations Using Special Relativity," *Am. J. Phys.* **54** (7), 631-636 (1986).

Kyrala, Ali, *Applied Functions of a Complex Variable* (Wiley-Interscience, New York, NY, 1972).

Laugwitz, Detlef. (tr. Fritz Steinhardt), *Differential and Riemannian Geometry* (Academic Press, New York, NY, 1965).

Lay, David, *Linear Algebra and Its Applications*, 4th ed. (Addison-Wesley, Reading, MA, 2011).

Leithold, Louis, *The Calculus, with Analytic Geometry*, 2nd ed. (Harper and Row, New York, NY, 1972).

Lichnerowicz, A., *Elements of Tensor Calculus* (Methuen and Co., London, UK, 1962).

Logan, John D., *Invariant Variational Principles* (Academic Press, New York, NY, 1977).

Marion, Jerry B., and Thornton, Stephen T. *Classical Dynamics of Particles and Systems*, 5th ed. (Brooks/Cole, Belmont, CA, 1994).

Mathews, Jon, and Walker, R. L., *Mathematical Methods of Physics* (W. A. Benjamin, Menlo Park, CA, 1970).

Merzbacher, Eugen, *Quantum Mechanics*, 2nd ed. (John Wiley and Sons, New York, NY, 1970).

Misner, Charles W., Thorne, Kip S., and Wheeler, John A., *Gravitation* (Freeman, San Francisco, CA, 1973).

Moore, John T., *Elements of Abstract Algebra* (Macmillan, London, UK, 1967).

Neuenschwander, D. E., *Emmy Noether's Wonderful Theorem* (Johns Hopkins University Press, Baltimore, MD, 2011).

Newton, Isaac, *The Principia*, tr. Andrew Motte (Prometheus Books, Amherst, NY, 1995).

Ohanian, Hans C., *Gravitation and Spacetime* (W. W. Norton, New York, NY, 1976).

Oprea, John, *Differential Geometry and Its Applications* (Mathematical Association of America, Washington, DC, 2007).

Panofsky, Wolfgang K. H., and Phillips, Melba, *Classical Electricity and Magnetism*, 2nd ed. (Addison-Wesley, Reading, MA, 1965).

Pauli, Wolfgang, *Theory of Relativity* (Dover Publications, New York, NY, 1981).

Peacock, John A., *Cosmological Physics* (Cambridge University Press, Cambridge, UK, 1999).

Peebles, P. J. E., *Principles of Physical Cosmology* (Princeton University Press, Princeton, NJ, 1993).

Rindler, Wolfgang, *Relativity: Special, General, and Cosmological* (Oxford University Press, Oxford, UK, 2001).

Roy, R. R., and Nigam, B. P., *Nuclear Physics: Theory and Experiment* (John Wiley and Sons, New York, NY 1967).

Sakurai, J. J., *Advanced Quantum Mechanics* (Addison-Wesley, Reading, MA, 1967).

Salam, Adbus, *Unification of Fundamental Forces: The First of the 1988 Dirac Memorial Lectures* (Cambridge University Press, Cambridge, UK, 1990).

Schouten, J. A., *Tensor Analysis for Physicists* (Clarendon Press, Oxford, London, UK, 1951).

Schutz, Bernard F., *A First Course in General Relativity* (Cambridge University Press, Cambridge, UK, 2005).

Schutz, Bernard F., *Gravity from the Ground Up* (Cambridge University Press, Cambridge, UK, 2005).

Smart, J. J., *Problems in Space and Time* (Macmillan Company, New York, NY, 1973).

Statchel, John, ed., *Einstein's Miraculous Year: Five Papers That Changed the Face of Physics* (Princeton University Press, Princeton, NJ, 1998).

Symon, Keith R., *Mechanics* (Addison-Wesley, Cambridge, MA, 1953).

Synge, J. L., and Schild, A., *Tensor Calculus* (Dover, New York, NY, 1978; University of Toronto Press, Toronto, ON, 1949).

Taylor, Angus E., and Mann, Robert W., *Advanced Calculus*, 2nd ed. (Xerox Publishing, Lexington, MA, 1972).

Taylor, Edwin F., and Wheeler, John A., *Exploring Black Holes: Introduction to General Relativity* (Addison Wesley Longman, San Francisco, CA, 2000).

Taylor, Edwin F., and Wheeler, John A., *Spacetime Physics* (Freeman, San Francisco, CA, 1966).

Taylor, John R., *Classical Mechanics* (University Science Books, Sausalito, CA, 2005).

Tolman, Richard C., *Relativity, Thermodynamics, and Cosmology* (Dover Publications, New York, NY, 1987; Oxford University Press, Oxford, UK, 1934).

Turner, Brian, and Neuenschwander, D. E., "Generalization of the Biot-Savart Law to Maxwell's Equations Using Special Relativity," *Am. J. Phys.* **60** (1), 35-38 (1992).

Vanderlinde, Jack, *Classical Electromagnetic Theory* (John Wiley and Sons, New York, NY, 1993).

Wardle, K. L., *Differential Geometry* (Dover Publications, New York, NY, 1965).

Weart, Spencer R., ed., *Selected Papers of Great American Physicists* (American Institute of Physics, New York, NY, 1976).

Weinberg, Steven, *Gravitation and Cosmology: Principles and Applications of the General Theory of Relativity* (Wiley, New York, NY, 1972).

Weinhold, Frank, "Thermodynamics and Geometry," *Physics Today,* March 1976, 23-30.

Weyl, Hermann, *Space, Time, Matter* (Dover Publications, New York, NY, 1952).

Yaglom, I. M., *Complex Numbers in Geometry* (Academic Press, New York, NY, 1968).

Index